# SPECIAL STUDIES

# IDEAS and WEAPONS

EXPLOITATION OF THE AERIAL WEAPON BY THE UNITED STATES

DURING WORLD WAR I; A STUDY IN THE RELATIONSHIP OF

TECHNOLOGICAL ADVANCE, MILITARY DOCTRINE, AND THE

DEVELOPMENT OF WEAPONS

## by I. B. HOLLEY, JR.

New Imprint by the
Office of Air Force History
Washington, D.C., 1983

ii

**Library of Congress Cataloging in Publication Data**

Holley, I. B. (Irving Brinton), 1919–
  Ideas and weapons.

  Reprint. Originally published: Yale University Press. 1953.
  Bibliography: p. 179.
  Includes index.
  1. Aeronautics, Military—United States. 2. World War—1914–1918—Aerial
operations, American. I. Title.
UG633.H6   1983                    358.4′00973                    83–18967
ISBN 0–912799–11–0

The views expressed in this publication are those of the author and should not be construed to represent the policies of the United States Air Force or the Department of Defense.

For sale by the Superintendent of Documents, U.S. Government Printing Office,
Washington, D.C. 20402

# *Foreword*

Few individuals have contributed more to the study of history in the United States Air Force than Professor I. B. Holley, Jr., of Duke University. After wartime service in the Army Air Forces, Holley joined the Air Force Reserve in 1947, rising to the rank of Major General in 1976. In a succession of positions, in the Office of the Secretary of the Air Force, as mobilization designee to the Commander of the Keesler Technical Training Center, and from 1975 to retirement in 1981 as mobilization designee to the Commander of Air University, Professor Holley strove for excellence in Air Force education. In numerous other ways—as a member of advisory panels and committees on ROTC and Air Force and NASA history, lecturing to Air University faculty and Air Staff research associates on research methods, training officers and other students for the Ph.D. in military history, presenting papers and speaking on military topics from leadership to space operations—he has for more than thirty-five years helped to give the nation's youngest service an appreciation for its rich and varied past.

From the beginning of his scholarly career, Holley has been concerned with the influence of thought on military organization and on war. *Ideas and Weapons* grew out of his studies at the end of World War II on the evolution of weapons in the period of the World Wars. It evolved into a dissertation at Yale University and has since become a classic of airpower history. In this work and in his later writings, Holley has emphasized the crucial role doctrine plays in air forces. In his Harmon Memorial Lecture at the U.S. Air Force Academy in 1974, Holley quoted an early definition of doctrine from the Joint Chiefs: "A compilation of principles . . . developed through experience or by theory, that represent the best available thought." Holley agreed that "doctrine is what is officially approved to be taught." But he knew it to be far more important, especially to airpower: "Basic doctrine defines the roles and missions of the service, the scope and potential capabilities of its weapons systems." It "lies behind the decisions as to what weapons will be developed and gives guidance as to the relative importance of several competing roles or weapon systems when the time arrives to apportion the invariably inadequate supply of dollars." "Doctrine is like a compass bearing; it gives us the general direction of our course." And: "Doctrine is the point of departure for virtually every activity in the air arm." Equal in importance to content, Holley argued, was the process in a military organization by which doctrine came to be formulated. Holley addressed this issue in a penetrating article, "The Doctrinal Process: Some Suggested Steps," in a 1979 issue of *Military Review*.

Between that article and *Ideas and Weapons* lay nearly forty years of reflection on the relationship between weapons procurement, doctrine, and success in war. The Office of Air Force History is reprinting this book in order to keep it available for the use of the Air Force and for the general public. In 1982 the Chief of Staff called on the Air Force to continue the "study of military history, combat leadership, the principles of war and, particularly, the applications of airpower." This classic work is perfectly appropriate, for Professor Holley undertook the "study . . . in an effort to distill from past experience in the development of air materiel those lessons which might be of help in formulating policies for exploiting the air weapon more successfully in the future."

Richard H. Kohn
Chief, Office of Air Force History

# *Introduction*

SINCE TIME IMMEMORIAL weapons have played a significant role in tipping the scales of victory from one side to another. The weapons of antiquity developed slowly, sometimes imperceptibly, from age to age and from generation to generation, but nonetheless the side with superior weapons almost inevitably emerged from every struggle victorious. From prehistoric ages down nearly to our own era, the pace at which weapons evolved was determined by custom rather than by any systematic and conscious series of decisions. In recent years the pace has accelerated. Advances in science and technology have made possible the atom bomb. But the degree to which scientific and technological advances are exploited for military purposes depends upon the methods devised to secure that end. The haphazard and unsystematic means of other ages have yielded to a more orderly process of conscious decision, development, test, and evaluation, but even so these methods have lagged behind the creative forces of science. The vastly enlarged destructive potential of today's weapons makes the process of selection of great moment. The survival of nations or even of whole cultures may depend upon the ability to procure superior weapons. It behooves us to be certain that our system is adequate to ensure this superiority. The experience of men who have grappled with this problem in the past should prove valuable to those who must deal with the question in the future.

# United States Air Force
# Historical Advisory Committee

(As of September 1, 1983)

Lt. Gen. Charles G. Cleveland,
USAF
*Commander,* Air University

Mr. DeWitt S. Copp
The National Volunteer Agency

Dr. Philip A. Crowl
Annapolis, Maryland

Dr. Warren W. Hassler, Jr.
Pennsylvania State University

Brig. Gen. Harris B. Hull, USAF,
Retired
National Aeronautics and Space
Administration

Dr. Alfred F. Hurley
Brig. Gen., USAF, Retired
North Texas State University

Mr. David E. Place
*The General Counsel,* USAF

Gen. Bryce Poe II, USAF, Retired
Alexandria, Virginia

Lt. Gen. Winfield W. Scott, Jr.
*Superintendent,* USAF Academy

Dr. David A. Shannon (Chairman)
University of Virginia

# Preface

THIS STUDY GREW out of the author's wartime experience at Wright Field, Dayton, Ohio. While there, during the last 18 months of World War II, the author wrote three separate monographs relating to the development of air materiel: "Evolution of the Liaison-Type Airplane, 1917–1944"; "Development of Aircraft Gun Turrets in the Army Air Forces"; and "Rotary-Wing Aircraft in the Army Air Forces, a Study in Research and Development Policies." As these studies took shape it became clear that all three, despite their entirely different subjects, were closely related. Each of the studies revealed a common theme: the pace of development for any weapon during the between-war years is chiefly determined by the extent to which its mission or operational function is known and defined. When there is no effective system for determining doctrine, the pace of development is necessarily slow.

The author's interest in the problem of the development of materiel led him, in cooperation with a group of others, to undertake a program to speed the process of translating ideas into weapons by improving existing procedures for handling experimental projects. The student of history seldom has an opportunity to put his findings to such immediate use, and most gratifying of all was the realization that the historical approach alone could offer the depth of perspective necessary to attack this contemporary problem in engineering.

A combination of the work involved in writing the three monographs and the applied history encountered in the project dealing with methods of engineering led to the present inquiry. This book, part of a much larger project, attempts to explore something of the background of the contemporary air weapon. The study was undertaken in an effort to distill from past experience in the development of air materiel those lessons which might be of help in formulating policies for exploiting the air weapon more successfully in the future. If the lessons derived are sound, it may be possible to find some principles for developing weapons in general.

*Acknowledgments*

So MANY PEOPLE have assisted in the preparation of this study that it is impossible to thank each one. The following individuals were especially helpful over a period of many months: at the Industrial College of the Armed Forces, Dr. M. S. Reichley, Miss Clara Widger, Dr. B. H. Williams, Col. A. E. Michelsen, S.C., Dr. L. C. Hunter, and CWO E. J. Sands; at the National War College, Miss K. E. Greenwood; and at Yale University, Professors R. E. Turner, L. W. Labaree, R. E. Gabriel, L. P. Curtis, and Sherman Kent all contributed generously with their time and experience. Miss Elizabeth B. Drewry smoothed the way at the National Archives and Lt. Col. J. D. Hamilton, C.E., helped carry the completed manuscript over the necessary military hurdles. Without the financial assistance provided by the Duke University Research Council this study would never have reached publication.

THE USAGE of the War Department Special Staff Historical Division requires all histories of World War II to include a note on footnotes, especially where archival material has been consulted. Although this study is not a portion of the World War II series, the practice is applicable here. Research for this inquiry involved extensive use of two bodies of archival materials and occasional use of several other collections. All abbreviations used in citing these collections are listed below in the glossary, and the bibliographical note at the end of the text gives a lengthy explanation of the source materials consulted. Nonetheless, further to assist the reader, the following sample citations are presented to illustrate the peculiarities of usage involved here.

The citation NA, BAP Hist. Box 30, 452.1 Airplanes, General means, at length, National Archives, Bureau of Aircraft Production Historical File, shelf box number 30, file folder catalogue number 452.1 (according to the decimal file system of the War Department) entitled Airplanes, General. Similarly, the citation NA, WWI Orgn. Records, A.S. Hist. Records Box 1 means, in full, National Archives, World War I Organization Records, Air Service Historical Records, shelf box number 1. The citation AAF Archive refers to items in the collection of files maintained by the Air Historical Office of the AAF. Actually there is no such official designation as AAF Archive, but in the absence of a descriptive phrase differentiating between the administrative files and the historical files of the Air Historical Office some such arbitrary description appeared necessary.

For the reader who is not familiar with the problems of citation encountered in publications of the War Department and Congress, as well as to simplify the problem of using citations in general, the practices mentioned below are singled out for special attention. Where titles are used in a number of chapters, the citation is repeated at length at its first appearance in each new chapter. Where multivolume Congressional hearings or other complicated documents are used, each citation is repeated in full. Such repetition may appear needlessly cumbersome, but avoiding the confusion which would result from different days of testimony, different sets of page numbers used in reprints, and other such irregularities makes full citation worth the trouble.

Congressional publications pose no end of problems with respect to citation. Style manuals, for the most part, wisely avoid the subject as if it were the plague. Common sense has been the determinant where precedent is unknown. Where Congressional documents are undated save by the traditional phrase "ordered to be printed," that date has been chosen. The author has occasionally employed brief forms in lieu of yard-long descriptive titles.

In citing publications of the War Department, it has not appeared necessary to add the obvious information: Government Printing Office (GPO), Washington, in addition to the date. All publications of the War Department, unless otherwise indicated, are printed by the GPO in Washington. The old army custom by which a superior officer puts his name to the work of juniors is annoying when encountered in academic circles, but there can be no escape from the practice where the chain of command prevails. Departmental publications generally have been credited to individuals where Library of Congress classification credits the individuals; otherwise publications are credited to the organization of origin.

The question of military abbreviations is always vexing. In such a study as this, prepared for both military and nonmilitary readers, the question is peculiarly difficult. Such a common term as AEF needs no explanation. On the other hand, lay readers might find OCSigO rather esoteric. With regrets for brevity, all such organizational designations have been dropped or, where used, explained at length. The glossary may be consulted should one find these precautions insufficient.

One last word remains to be said on citation. Manuscript studies, e.g., typescript drafts of chapters in the Bureau of Aircraft Production history, are cited in quotation marks to differentiate them from correspondence, military reports, staff studies, and other similar administrative material. Where the writer of these manuscript studies is unknown, the reader may reasonably assume that he was a member of the organization mentioned in the portion of the citation following the subject matter. Where, in the preparation of this study, the author consulted typescript drafts of chapters such as the one mentioned above and, in addition, checked the contents of the draft against the working papers from which it was written, these papers are mentioned as "research data" in the accompanying citation. The bibliographical note discusses this problem in greater detail.

## GLOSSARY

| | |
|---|---|
| Act. | Acting |
| Admin. | Administrative Office |
| AFCF | Air Force Central Files |
| AGO | Adjutant General's Office |
| A.S. | Air Service |
| Asst. | Assistant |
| BAP | Bureau of Aircraft Production |
| Com. | Committee |
| Cong. | Congress |
| Dir. | Director |
| Div. | Division |
| DMA | Division of Military Aeronautics |
| Doc. | Document |
| Equip. | Equipment |
| Exec. | Executive Office |
| F.Y. | Fiscal Year |
| Hist. | History |
| Ind. | Indorsement |
| Info. | Information |
| JANTB | Joint Army-Navy Technical Board |
| MID | Military Intelligence Division |
| NA | National Archives |
| NACA | National Advisory Committee for Aeronautics |
| NRC | National Research Council |
| Orgn. | Organization |
| OCAS | Office, Chief of Air Service |
| Rpt. | Report |

| | |
|---|---|
| Sen. | Senate |
| Sect. | Section |
| Sess. | Session |
| TAG | The Adjutant General |
| Tech. | Technical |
| TICAF | The Industrial College of the Armed Forces |
| WF | Wright Field |
| WWI | World War I |
| WFCF | Wright Field Central Files |

# Contents

TO CHARLES M. THOMAS

*An able officer, a kind teacher, a warm friend,
who told me not to write this book*

*PART ONE*: *The Problem Presented*

# Chapter I. The Development of Weapons: Procedures and Doctrine

An English army of several thousand men led by a renowned officer landed on the coast of Normandy and pressed eastward in a raiding expedition penetrating almost to Paris. The officer was Edward III, king of England, the time, July 1346. However remote the day, now more than 600 years past, the expedition is still worthy of study for its military lessons.

Edward's troopers loitered and plundered along the way until they were suddenly confronted with a French host hurriedly gathered to resist their advance. The spot was not a strategic one for battle, and since the fleet of convoys which carried the English army across the Channel had returned home, retreat along the path of advance was impossible. The only alternative to fighting was withdrawal toward Flanders. Crossing the Seine near Paris, the English made for the Somme, but here they found the crossings guarded as they tried the fords one after another down the length of the river. At last, with some difficulty the whole English force managed to slip across the salt flats below Abbéville just ahead of the flood tide which prevented French pursuit for a full 12 hours. With the period of grace thus secured, Edward led his troops through the forest of Crécy and at leisure selected a defensive position with the wood at his back and a long, gentle, downward slope of open ground before him. Here, on a site of his own choosing, the king drew up his men in battle array, three great blocks or *batailles* of dismounted knights and men-at-arms with connecting ranks of archers armed with English longbows.

The French forces under King Philip of Valois approached this position in a disorder which reflected both the speed of pursuit and the confusion of a hurried river crossing. Against the French king's wishes, the rash and undisciplined feudal lords assailed the English position. Each new group of Frenchmen to arrive on the scene thrust forward in attack, and without exception each suffered the same fate; the archers with English longbows stopped the drive before the French could fairly engage the standing men-at-arms and dismounted knights.

Medieval chronicles are notoriously unreliable when dealing with numbers, but even if one rejects Froissart's figures, the evidence still indicates that the English won the Battle of Crécy with a force approximately half that of the French. And with so markedly inferior a force the English archers ended the long supremacy of feudal cavalry. If the French dead in this battle (more than 1,500 "lords and knights" on the field of Crécy) were not enough to spell out the revolution achieved by the longbow, later events in the Hundred Years' War, when the French learned to dread the English arrow, made the implications of the new weapon only too clear.[1] Sir Charles Oman says the fight at Crécy was "a revelation to the Western World," a startling demonstration of the supremacy of the longbow over the armored knight on horseback.

One would assume that the English kings must have been seeking eagerly to counterbalance their country's inevitable numerical inferiority with such a weapon as had wrought this revolution in arms. On the contrary, the longbow appears to have been on the English back doorstep for nearly 250 years before Crécy. English warfare from the time of the Norman invasion to Edward I—1066 to 1277—was of two sorts: Continental wars in which mailed horsemen did the principal fighting and infantry were of little concern and local wars with the Irish and Welsh. A Welsh historian, Giraldus Cambrensis, whose *Expugnatio* appeared sometime in the middle of the 12th century, wrote at length on the Welsh use of the longbow in the border wars and recommended an increase in the number of Welsh archers in the Anglo-Norman armies to enhance their firepower.[2] Like the advice of many military historians, this proposal appears to have gone unheeded.

The bow, of course, had long been known in England. Archers armed with the short bow, known since Roman times, had participated in the fray at Hastings. But the insignificance of the bow is revealed by the absence of any mention of it in the Assize of Arms held by Henry II in 1181. By the time of the next assize in 1252 during the reign of Henry III, the influence of the Welsh had become apparent; citizens with 40-shilling holdings or less were required to appear at the muster armed with the longbow. During the Welsh and Scottish border wars of the 13th century,

---

1. This account is based upon Charles Oman, *A History of the Art of War in the Middle Ages* (London, Methuen, 1898), pp. 597–615.
   2. *Ibid.*, p. 400.

Edward I perfected the use of the longbow in conjunction with cavalry until finally in 1298 at the Battle of Falkirk the English, using longbows, demolished a Scots force under Wallace. Unfortunately, English chroniclers in recording the battle "forgot that the archers had prepared the way, and only remembered the victorious charge of the knights at the end of the day." [3] The importance of comprehensive tactical analysis was unrecognized, and Edward's lesson was lost when the king died in 1306 without leaving a written record of his military art. When the Scots under Robert Bruce put Edward's son and his English army to rout at the Battle of Bannockburn in 1314 by using a judicious combination of cavalry and longbowmen, Bruce proved himself the abler pupil of Edward I. The training acquired in continual border wars enabled Edward III to lead to Crécy an army skilled in the use of the longbow, which worked such havoc among the "fiery and undisciplined noblesse" of the French.

Oman finds it "rather surprising" that Edward III was so slow in heeding the "obvious" lesson of the preponderant influence of the longbow and increasing the proportion of bowmen in his forces.[4] How much more surprising is the painfully slow advance of the longbow as an English weapon. There are nearly 250 years between Giraldus Cambrensis' advocacy of the Welsh elm bow and Crécy, yet the lesson of the border wars was plain: a new weapon gave one side an advantage over the other. Crécy is chosen to illustrate this principle because the battle took place over 600 years ago and is sufficiently remote to be free from all interests, prejudices, and emotions which surround so many present military practices. Other examples are plentiful. In 479 B.C. at the Battle of Plataea the Persian rabble fled in dismay before Greeks using an innovation in warfare, which consisted of a phalanx of troops marching in step with shields aligned—in truth, a mobile armored force.[5] In the spring of 1940 a handful of British fighters broke the back of the German aerial invasion because they had an innovation called radar.

Sometimes the advantage of a superior weapon is decisive before countermeasures can be evolved. It follows then that the methods used to select and develop new weapons and the doctrines concerning their use will have an important bearing upon the success or

3. *Ibid.*, p. 569.
4. *Ibid.*, 2, 57–124.
5. T. Wintringham, *The Story of Weapons and Tactics from Troy to Stalingrad* (Boston, Houghton Mifflin, 1943), p. 29.

failure of armies—and of nations. A brief résumé of some of the more important developments in weapons and the circumstances surrounding their adoption in the United States should provide an adequate perspective for the narrower problem which is the subject of this study. One need go no further back than the 19th century.

Consider, for instance, the annual report of Joel R. Poinsett, Secretary of War in 1840, which discussed at length the problem of introducing new weapons. After reviewing a number of projects undertaken by the Ordnance Department the secretary declared that the necessities of national security generally inclined him "to discountenance" all "new inventions" unless convinced of their superiority "by long-tried experiments in the field." In the matter of breech-loading weapons, the secretary was emphatic: "I fear that every attempt . . . will fail as they have hitherto done, after involving the government in great expense." On the other hand, the percussion cap for flintlock muskets found official favor inasmuch as this particular innovation had been "fairly tested in the field by the armies of Europe." [6] The policy of the War Department, it appears, was to follow, not lead. When a patent breech-loading carbine was offered to the department in 1842, the Colonel of Ordnance agreed to a trial of the new weapon but noted that it was not customary for the government to incur any expense beyond the consumption of ammunition. The colonel was quick to point out that not all the fault lay with Ordnance: "A prejudice against all arms loading at the breech is prevalent among officers, and especially the Dragoons." Moreover, the colonel doubted that the new breechloader could be introduced into the service even if it were found to be better than other models.[7]

Between 1842 and 1845 the Ordnance Department conducted a number of tests on breech-loading weapons. The results were extremely discouraging, as might be expected of a new technological process in the testing stage. The Colonel of Ordnance, an official whose status might be fairly translated as Chief of Ordnance in later times, reported on the problem to the Secretary of War:

Upon due consideration of the subject the department decided on abandoning the manufacture of breech-loading

6. Brig. Gen. S. V. Benét ed., *Annual Reports and Other Important Papers Relating to the Ordnance Department* (Washington, Government Printing Office [hereafter GPO], 1878), *1*, 381–2; hereafter cited as *Ordnance Reports*. See below, bibliographical note.
7. *Ibid., 1*, 435–6.

arms, and have followed in the steps of the great powers of Europe, deciding that a diversity of arms was productive of evil, and adopting those of ordinary construction which are the simplest and easiest managed by the common soldier.

So firmly convinced of the virtues of muzzle-loading muskets of "ordinary construction" was this colorful colonel that he put himself on record concerning the soon-to-be-famous Colt's patent arms: "That they will ultimately all pass into oblivion cannot be doubted . . ." Meanwhile, he warned, it would be well for officers to take care not to be "ensnared again by the projects of inventors." [8] The patent carbine which the Colonel of Ordnance found so undesirable was able to fire more than 14,000 rounds before it broke down in proving trials. Unfortunately a service test with troops in the field was hard to obtain. The company officer to whom the carbines were issued must have been a dragoon; he replied, when pressed for a report on service tests, that the carbines were not worth the storeroom they occupied.[9] A hundred years later, procedures for following up service tests were still a troublesome matter.

When summarizing the whole problem of breechloaders in 1851, a subsequent Chief of Ordnance made it clear that his department was not utterly blind to the innovation. He admitted the real advantage of breech-loading weapons but indicated that these advantages were difficult if not impossible to obtain without sacrificing the essential qualities of simplicity and durability. The department would continue to use muzzle-loaders until it encountered "convincing proof" of superior breechloaders.[10] Here then was a step forward. The Chief of Ordnance recognized the validity of the principle of breech-loading and differentiated between the principle as an objective sought and individual inventions which failed, for mechanical or technical reasons, to satisfy the requirements of the principle. His view represented a marked advance over the attitude of the previous Colonel of Ordnance who had summarily rejected the principle of breech-loading weapons merely because repeated attempts at application had ended in failure.

By 1859 war and the rumor of war had worked a real change in the Ordnance Department, which now professed to "encourage the application of scientific knowledge and mechanical skill to im-

8. *Ibid., 2*, 3–4.
9. *Ibid., 2*, 8–9.
10. *Ibid., 2*, 381.

provements in arms." The department was on the verge of adopting a breech-loading carbine, but "uniformity of armament" was so obviously essential for training and for supply of ammunition in time of war that Ordnance officials were reluctant to issue any one new type before deciding upon the best. The final selection, it was pointed out, might well fall upon an arm "not yet invented." [11] This desire for the utmost qualitative superiority was admirable, but with open rebellion a few months off, the time for decisions on what to produce was already at hand, even if it was a weapon somewhat short of the ideal. Secretary of War John B. Floyd was certainly not unaware of the potentialities of breechloaders. After reviewing the "wonderfully numerous" experiments with the innovation, he considered the best of the weapons "by far the most efficient arms ever put into the hands of intelligent men" and recommended that immediate steps be taken to arm all light troops with breechloaders. To do less, he declared, was "an inhuman economy." [12] Unfortunately for the Federal cause, in this respect at least, Secretary Floyd "went South," and the Ordnance Department continued to seek the best breechloaders but not to issue them.

As late as February 1861 the Colonel of Ordnance declared that the muzzle-loader of the service was "unsurpassed for military purposes." And the value of repeating arms was curtly dismissed by the colonel, who pointed out that they had been known to misfire and that front-rank men would be "more in dread of those behind than of the enemy." That repeating arms would do away with the tactical maneuver of multiple ranks attacking in close order across open ground seems never to have occurred to this officer. His was by no means an isolated expression of opinion. At about the same time another Ordnance officer said of the musket issued by the United States that there is "no superior arm in the world," an opinion he was willing to back by proposing that the Ordnance Department absolutely refuse to answer any requisitions for new and untried arms.[13]

By 1864 the pressure of wartime operations had changed a great many opinions and replaced several key officers in the Ordnance Department. The new officials accelerated the pace of experiment, and both breechloaders and repeating arms were issued in relatively small quantities to the troops in the field for service tests. While

11. *Ibid., 2,* 669.
12. Secretary of War, *Annual Report,* 1860, Sen. Doc., 36 Cong. 2 Sess., Vol. *2.*
13. *Ordnance Reports, 4,* 842–5.

lamenting the difficulties of securing accurate reports from the troops on the merits or demerits of any given weapon, the new Chief of Ordnance reported that urgent demands from the field left no doubt that repeating arms were the favorite of the army.[14] Similarly, Secretary of War E. M. Stanton was informed that breech-loading weapons were now "greatly superior" to the musket manufactured by the national armories and that the time had come to decide upon a breechloader for issue to the service. The moment was in December 1865.[15]

To assume that the adoption of breechloaders must be a simple matter, once official opinion lined up behind the project, would be naïve. There is a great difference between the giving of an order and its actual execution in every detail. In 1867 when the war was safely in the background, a Joint Congressional Committee on Ordnance presented a resolution to stop the modification of service muskets into breechloaders since such modification would render useless existing stocks of ammunition.[16] Reasons of economy no doubt motivated this Congressional interference in a technical decision. Congress might well have interfered sooner, for there were more than a million obsolete muskets unissued at the end of the war.[17]

The well-known British military critic and student of war, Liddell Hart, in commenting on the Union and Confederate armies, credits the few repeating arms which actually reached the hands of Federal troops in action with a "decisive influence" out of all proportion to their numbers. He bolsters his contention by quoting Confederate Gen. E. P. Alexander to the effect that the war might have been terminated within one year if the Federal infantry had been equipped with even the imperfect repeaters of 1861 design.[18] Liddell Hart makes a point of exceptional importance. In spite of the high quality of generalship exhibited in the war, armament lagged "well behind the pace of invention." But more important than the hither edge of invention, which sometimes lies beyond the

14. *Ibid.*, *4*, 882–3.

15. *Ibid.*, *4*, 893–4. See also F. A. Shannon, *The Organization and Administration of the Union Army; 1861–1865* (Cleveland, Arthur H. Clark, 1928), p. 142.

16. *Ordnance Reports*, *4*, 903.

17. Shannon, *Union Army*, p. 123, from *War of the Rebellion; compilation of official records of Union and Confederate Armies* (Washington, GPO, 1880–1901), Ser. 3, *5*, 145.

18. B. H. Liddell Hart, *The British Way in Warfare* (London, Faber and Faber, 1932), pp. 121–2. See also E. P. Alexander, *Military Memoirs of a Confederate* (New York, Scribner's, 1908).

scope of production, was the failure to utilize new weapons which were not only technically possible but capable of being produced on an extensive scale.

Shannon, the author of a classic study on the Union Army, makes the same point even more forcibly. The North, with its control of the seas and adequate industry, was free to choose the weapons it wanted. Unfortunately, the North's choice was not the best weapon available but a musket modified since the Revolution by little more than the addition of the percussion lock and rifling. Thus the North fought with the same weapons available to the South and made slight use of the superior arms within its grasp. Shannon considers it a strange paradox that the North used every means, including bribery, to increase its firepower by pressing more and more men into the ranks and at the same time failed, until late in the war, to increase firepower by putting better weapons in the hands of the trained men already in the ranks.[19] Eighty-odd years later the problem of correlating technological advance in weapons with higher national policy was still far from being entirely solved.

If armies have been slow in applying the maxim that superior arms favor victory, it may be shown that their intransigeance has resulted to a great extent from three specific shortcomings in the procedure for developing new weapons. These shortcomings appear to have been: a failure to adopt, actively and positively, the thesis that superior arms favor victory; a failure to recognize the importance of establishing a doctrine regarding the use of weapons; and a failure to devise effective techniques for recognizing and evaluating potential weapons in the advances of science and technology.

Although military men have been slow to recognize and put into practice the thesis that superior arms favor victory, military writers down through the ages have given some recognition to the importance of weapons. Vegetius in his *Military Institutes*, which has been aptly described as the field service manual of the Imperial Roman Army, recognized the relative importance of materiel. "The Legion," said Vegetius, "owes its success to its arms and machines, as well as to the number and bravery of its soldiers."[20] Authorities can be found repeating the truism in every century

19. Shannon, *Union Army*, pp. 108–9, 140. For pertinent comment on this problem in World War II, see J. P. Baxter III, *Scientists against Time* (Boston, Little, Brown, 1946), chap. ii.

20. Vegetius, *Military Institutes*, tr. John Clarke (London, 1767), Bk. II, sect. xxv.

down to our own, pointing out that fighting soon led men to special inventions which they turned to their advantage.[21] Yet a decided disparity has prevailed between frequent assertion of the thesis that inventions could be put to military advantage and the paucity of studies on the application of the thesis in practical terms. Most military writers have bowed obsequiously in the general direction of the principle, but having done this they rush on to the supposedly more important subjects of strategy and command. Sometimes tactics are treated with attention to detail, but weapons have generally been dismissed with the slighting treatment combat soldiers usually reserve for rear-echelon supply troops. A bare handful of writers has specialized in the problem of weapons; almost all others, dealing more generally with warfare, have either relegated the subject to a minor position or ignored it entirely.[22]

The great Karl von Clausewitz, who dominated military thinking for nearly a hundred years following the demise of the Napoleonic Empire, admitted in *Vom Krieg* that superiority in the organization and equipment of an army has at times given "a great moral preponderance," but having made this concession he points out how clear it must be that "arming and equipping are not essential to the conception of fighting." Even while conceding that fighting determined the character of arms and that arms modified the character of war, Clausewitz restricted the "art" of war, by entirely arbitrary definition, to the actual conduct of battle. To include the problems of arms and equipment, he said, would be to establish a special case rather than a timeless principle.[23] Unlike Adam Smith, Clausewitz wrote after the Industrial Revolution was well under way, but his writings show an utter lack of appreciation of the implications for the development of weapons in the new mechanization. Just how static Clausewitz' concept of the evolution of weapons was is shown in his contention that "completing

---

21. See, for example, Karl von Clausewitz, quoted in J. F. C. Fuller, *Armament and History* (New York, Scribner's, 1945), p. 1. See also J. F. C. Fuller, *The Foundations of the Science of War* (London, Hutchinson, 1926), p. 146, in which he quotes Carlyle's *Sartor Resartus,* chap. v: Man, "without tools he is nothing, with tools he is all."

22. A survey of three representative military collections—the War Department Library, the National War College Library, and the Library of the American Military Institute—gives ample evidence of the comparative neglect of the problem by military writers until very recent years. Probably the most prolific but not necessarily the most influential of the few students of the importance of armament is Brig. Gen. J. F. C. Fuller.

23. Carl von Clausewitz, *On War,* tr. J. J. Graham from German 3d ed. (London, 1873), *2,* Bk. II, chap. i, 43, and *3,* Bk. V, chap. iii, 3.

and replacing articles of arms and equipment . . . takes place only periodically, and therefore seldom affects strategic plans." [24] Weapons, it would appear, were taken for granted by the military theorists of the 19th century. Baron de Jomini, who was perhaps the leading military theorist of that century next to Clausewitz, made a concession to materiel in granting extensive consideration to logistics; yet even he defined logistics in the former sense of "the practical art of moving armies" rather than in the broader contemporary definition which embraces more of the element of production.[25]

The myopia demonstrated by these theorists had a profound influence upon those military leaders who dominated the profession of arms down to World War I. General Foch, when he published his *Principles of War* in 1903, carried on in the tradition of Clausewitz. His "principles" concerned strategy and tactics or the use of weapons. The selection, development, and procurement of superior weapons he ignored or assumed. When Foch at length came to dominate the councils of both France and the Allies, his emphasis on personnel rather than materiel helped determine the character of the armies that fought in World War I.[26] The absurdities created by the failure to emphasize the importance of superiority in weapons in the years leading up to 1914 were nowhere more vividly portrayed than in France. In the nation of the mass army Gen. F. G. Herr reported the prevailing attitude: "The battle will be primarily a struggle between two infantries, where victory will rest with the large battalions; the army must be an army of personnel and not of materiel." [27] This attitude probably marked the apogee of neglect for the thesis of superior weapons.

The events of World War I abruptly focused attention upon the relative significance of materiel in securing victory. The Italian general Douhet, philosopher of air power, expressed the new em-

24. *Ibid., 1,* Bk. II, 130. See also Karl von Clausewitz, *Principles of War,* tr. H. W. Gatzke (Harrisburg, Military Service Publishing, 1942), an attempt to distill "timeless principles," which is the more pointedly futile for its failure to embrace the potentialities of the development of weapons.

25. Baron A. H. Jomini, *The Art of War,* tr. W. P. Craighill and G. H. Mendell (London, 1862). See also *Logistics,* Vol. *1,* No. 1 (Oct. 1945).

26. Liddell Hart, *The British Way in Warfare,* pp. 11, 93. See also F. Foch, *The Principles of War,* tr. J. de Morinni from 1903 French ed. (New York, H. K. Fly, 1918). For an example of Foch's influence in favoring manpower over materiel, see Foch-Pershing cable of 23 June 1918, cited in J. J. Pershing, *My Experiences in the World War* (2 vols.; New York, Frederick A. Stokes, 1931), *2,* 123.

27. Gen. F. G. Herr, *L'Artillerie,* pp. 4–5, quoted by Fuller in *The Foundations of the Science of War,* p. 29.

phasis on materiel when he said, "The form of any war . . . depends upon the technical means of war available." [28] Douhet was, of course, a theorist whose writings could be said to represent little more than his own personal opinions; the writings of the Americans, Crowell and Crozier, on the new importance of industry and weapons in modern warfare were more significant as indices of the new trend.[29] World War I awakened in military circles a new realization of the importance of weapons, but the emphasis was on quantity rather than quality. The first postwar report by the Secretary of War emphasized the need for a broader scope of training for military men. New weapons and new methods of warfare, the secretary said, made it "specially apparent" that staff officers should have not only a wider knowledge of their purely military duties but also a "full comprehension of all agencies, governmental as well as industrial, necessarily involved in a nation at war." [30] This new awareness of the importance of industry received positive expression in the postwar provisions made for the planning of industrial mobilization and in the formation of the Army Industrial College to train officers in its techniques.[31] But materiel alone did not signify superiority of weapons: planning for industrial mobilization emphasized quantitative procurement, more weapons rather than better weapons. To be sure, centers for research and development and the millions devoted to improving weapons during this period show that the concept of superior weapons was not entirely neglected between the two world wars. Nevertheless, it was not until World War II and the approach of total war that military men and governments generally accepted and implemented the thesis of superior weapons as a cardinal tenet of military policy.[32]

28. G. Douhet, *The Command of the Air,* tr. Dino Ferrari from 1921 Italian ed. (New York, Coward McCann, 1942), Bk. I, chap. i, p. 6.

29. Benedict Crowell, Asst. Secretary of War, *America's Munitions; 1917–1918* (Washington, GPO, 1919), and William Crozier, Chief of Ordnance, *Ordnance and the World War* (New York, Scribner's, 1920). Both of these authors spoke with semiofficial authority, for they wrote from the records and experiences of the war in which they each had had important roles with regard to materiel.

30. Report of Secretary of War, *Annual Reports of the War Department,* 1919, *1,* 28.

31. J. M. Scammell, "A History of the Army Industrial College" (MS history, TICAF Library, 1947), *passim.*

32. The term "total war" is a generalization frequently abused. During World War II approximately 59% of industrial production in the United States (1942–44) was devoted to war purposes. See U. S. Department of Commerce, *Survey of Current Business* (Feb. 1946), p. 13. The belated formation in 1943 of the New Developments Division of the War Department General Staff after considerable

To carry the résumé of changing attitudes toward the thesis that superior weapons favor victory down through World War II would be to go beyond the scope of this study. The brief review already presented is useful, nevertheless, in that it makes more understandable the comparative paucity of interest and attention which military men have until recently devoted to the problem of revising doctrine to embrace new weapons. Without a tradition of positive and active adherence to this thesis as a prior condition, it is not surprising that the problem of relating doctrine to technological advance in weapons received only belated attention, in most instances long after the weapon itself had become available.

Superiority in weapons stems not only from a selection of the best ideas from advancing technology but also from a system which relates the ideas selected with a doctrine or concept of their tactical or strategic application, which is to say the accepted concept of the mission to be performed by any given weapon. Protracted and serious delays in the adoption of superior weapons have led critics to charge military men with congenital conservatism.[33] But it has sometimes happened that new weapons have been developed, adopted as standard, issued, and then neglected for lack of accepted doctrine regarding their use. It has probably more often happened that new weapons have been adopted and even used to a certain extent but that their full potential value has remained unexploited because higher policy-making echelons have failed to modify prevailing doctrine to embrace the innovation. New weapons when not accompanied by correspondingly new adjustments in doctrine are just so many external accretions on the body of an army.

Liddell Hart cites the case of Capt. Émile Mayer of the French army. A contemporary of Foch and Joffre at the École Polytechnique, Mayer accepted a position as military editor for the *Revue scientifique* where he became aware of the impact of military invention on doctrine. His prolific writings developed the thesis that new ideas—smokeless powder, for example—demanded new doctrines of war. Unfortunately, the revised doctrines he advocated

---

civilian pressure had been exerted is but one example of the lack of military emphasis on superior weapons. Another sign of the comparative neglect of superior weapons may be seen in the curriculum of the Army Industrial College, which did not emphasize the critical importance of research and development until *after* World War II.

33. See, for example, Brig. Gen. E. McFarland, Asst. Chief of Ordnance, "Trend in Weapons Types and Design," *Journal of the Franklin Institute, 230,* No. 4 (Oct. 1940), 415.

did not jibe with prevailing French military policy. Mayer was retired as a captain long before his contemporaries who were more willing to conform to accepted doctrines.[34] The incident is noteworthy only insofar as it serves to emphasize the difficulties involved in attempting to modify existing military thought. To introduce radical changes in the doctrines of warfare is to run headlong into the opposition of the entrenched interests. The bowyers' and fletchers' guilds were probably mortal enemies of the advocates of gunpowder. The belated demise of cavalry in the United States during 1946 and the anachronistic survival of captive balloons for the purpose of observation until the eve of World War II give some indication of the obstinate resistance of military institutions to doctrinal changes. But for all of this, the greatest stumbling block to the revision of doctrine was probably not so much vested interests as the absence of a system for analyzing new weapons and their relation to prevailing concepts of utilizing weapons.

"Victory smiles upon those who anticipate changes in the character of war," Douhet wrote, "not upon those who wait to adapt themselves after the changes occur." [35] Unfortunately, military men have had difficulty in providing the means of anticipating changes. General Fuller, one of the most prolific of British writers on warfare, may be unduly harsh when he says "soldiers are mostly alchemists," but he is probably correct in attributing the difficulty to a lack of scientific method in analyzing the elements comprising the revolutionary changes which have modified the character of warfare.[36] To go further into the reasons why armies have been slow in adjusting doctrine to advances in weapons would be to digress needlessly. Here it is important only to recognize the implications of this shortcoming. The events surrounding the development of doctrine for three well-known weapons will serve to illustrate the point that to adopt a new weapon without a new doctrine is to throw away advantage.

The machine gun was no new invention in 1914. As early as 1885 the modern machine gun was known in the United States. Even though the weapon had not yet emerged from the experimental stage, the Chief of Ordnance predicted then that it would in the future become "a prominent factor in every contest." [37]

34. Liddell Hart, *The British Way in Warfare*, p. 49.

35. Douhet, *The Command of the Air*, Bk. I, p. 30. Liddell Hart, *The British Way in Warfare*, p. 121, has almost the same thought.

36. Fuller, *The Foundations of the Science of War*, pp. 23, 31.

37. *Ordnance Reports, 3*, 190. See above, p. 6 n. 6.

Some years later, during the Russo-Japanese War of 1905, British observers reported that machine guns were working a "great execution." [38] But the experience of the Russo-Japanese War had no influence on British military doctrine as far as machine guns were concerned. Before the observers reported on the startling effectiveness of the novel weapon in actual warfare there were 24 machine guns in each British division or two per battalion. In 1914 the machine-gun strength of each division was exactly what it had been in 1899. In view of the scale of expenditures for other types of weapons during this period, it must certainly have been military policy and not limited appropriations which determined the number of machine guns authorized. By the end of 1918 there were over 500 machine guns in each British division. [39] The increase represented a revolution in concept, in doctrine, not a technological development.

Technical advances, to be sure, appeared in the machine gun during the period of World War I, but these were improvements and modifications rather than basic changes. The increased number of machine guns in each British division represented an advance in doctrine carried out at tremendous cost in blood. Even when prompted by mounting casualities, revision of the conventional doctrine was not easy. As late as 1915 one British commander considered the machine gun "a much over-rated weapon." Moreover, despite frequent German demonstrations of the machine gun's value, he felt that two per battalion were "more than sufficient." [40] On the other hand, Brig. Gen. C. T. Baker-Carr, a British officer who played one of the leading roles in revising doctrine on machine guns, probably recognized the real nature of the problem. He saw the delay in modifying military doctrine to fit the requirements of the new weapon as "the fault of the system" rather than "the fault of the individual." Baker-Carr possibly came even closer to the heart of the matter when he said, "The chief trouble at GHQ was that there was no one there who had time to listen to any new idea." [41] His observation is all the more revealing in that it echoes a sentiment expressed by Sir Percy Scott, "the Admiral Sims of

38. *Reports from British Field Officers Attached to the Japanese and Russian Forces in the Field, 2, 56*, cited in Fuller, *The Foundations of the Science of War*, p. 22. See also War Department General Staff, *Reports of Military Observers*, MID Report No. 8, Pt. 5, March 1907.

39. Col. J. F. C. Fuller, *The Reformation of War* (New York, E. P. Dutton, 1923), p. 86.

40. Brig. Gen. C. D. Baker-Carr, *From Chauffeur to Brigadier* (London, E. Benn, 1930), p. 87.

41. *Ibid.*, p. 89.

the Royal Navy." Admiral Scott considered the blindness of the Admiralty to new ideas a direct result of the failure of "administrative machinery" to provide "time to think of the needs of the future and how they should be met." [42] For want of "time to think" and for lack of an organization specifically charged with the function of relating doctrine to advances in weapons, the machine gun, although a standard item of equipment in 1914, was not fully exploited until well into the middle of World War I.

The tank, like the machine gun, came into prominence during World War I, but unlike the machine gun it evolved almost entirely within the war years. Interesting and pertinent though they may be, the details of the process by which the War Office (and the Admiralty, for that matter) were led to consider the idea of an armored tractor and develop it as a weapon lie somewhat beyond the horizon of this study. Nonetheless, the history of the tank, once it was produced in quantity and utilized in combat, closely parallels that of the machine gun. It might well be argued that from the battle of the Somme in September 1916 until Cambrai in November 1917 the tank was in the stage of proof testing. But the reduction in casualties and the ground gained when tanks were used thereafter conclusively showed the new weapon to be a revolutionary contribution to warfare. It is true that at the end of 1917 the tank still had far to go, but it had reached a point where even as an imperfect and faulty mechanism it was capable of exerting a significant influence in battle. Even so, in April 1918 the Royal Tank Corps was reduced from 18 to 12 battalions because infantry reinforcements were falling short.[43] In the crisis British military leaders clung to accepted doctrine; they favored manpower over materiel in securing victory. And even after the crisis had passed and while there was "time to think," official opinion continued to favor traditional concepts. The infantry was still considered "the arm which in the end wins battles," and the rifle and bayonet were thought to be the infantryman's "chief weapons." [44]

The same thought echoed officially in the United States, al-

42. Adm. Sir Percy Scott, *Fifty Years in the Royal Navy* (New York, Geo. H. Doran, 1919), pp. vi–vii. For a similar criticism of the administration of the War Office in Kitchener's time, see Graham Wallas, *The Art of Thought* (London, J. Cape, 1926), p. 137.

43. Fuller, *The Reformation of War*, p. 116, and *Armament and History*, p. 140.

44. British Army *Field Service Regulations*, 1924, quoted by Fuller in *The Foundations of the Science of War*, p. 30.

though the Surgeon General's statistics gave some evidence that the rifle and bayonet may not have been so important after all.[45] Military doctrine was slow to embrace the full implication of the tank. "I laugh at ideas," Marshal Foch is reputed to have said. "However good they may be, they possess value only insofar as they are translated into facts." [46] The tank was an idea; it had been translated into fact; yet its full value went unrecognized at the end of the war. Wars, it would appear, are governed not by the development of weapons but by such fractions of that development as have been recognized and incorporated into approved military doctrine.

The introduction of gas warfare presents a case somewhat similar to that of the tank. Two German scientists, Walther Nernst of the University of Berlin and Fritz Haber of the Kaiser Wilhelm Physical Institute, worked out the details of production and application of poisonous gas for use in the field. Then on 22 April 1915, at a point somewhat north of Ypres where the French and British lines joined, the Germans released a gas attack along a five-mile front. The results were staggering. After a 15-minute attack some 15,000 troops were thrown into confusion, and a great breach opened in the Allied lines. The British and French forces managed to close the breach but only after suffering 5,000 casualties and the loss of 60 field guns as well as other stores and equipment. In a war of position where every significant advance necessarily involved breaching the enemy's line as a preliminary condition, the gas attack at Ypres presented the German forces with an amazing opportunity. That they did not exploit the advantage resulted directly from a failure of the high command to adjust doctrine so as to meet the potential of the new weapon. But subsequent notable successes with gas—for example, the defeat of the British Fifth Army in March 1918—showed that the German high command was not always slow to learn from its own mistakes.[47] Statistics strengthen significantly the impression that the enemy in World War I recognized the full importance of relating doctrine with novel weapons. Figures compiled by the Surgeon

45. Report of Chief of Staff, *Annual Reports of the War Department*, 1920, Vol. *1*. Gunshot and bayonet wounds show a comparatively low incidence in relation to casualties from other causes, e.g. gas.

46. Quoted by Maj. Gen. E. D. Swinton, British advocate of the tank, in *Eyewitness* (London, Hodder and Stoughton, 1932), p. 80.

47. Brig. Gen. A. H. Waitt, *Gas Warfare* (New York, Duell, Sloan and Pearce, 1942), p. 21.

General in the United States demonstrate that 27.3 per cent of the casualties suffered by the AEF were from gas.[48]

In brief historical sketches the pages above have shown that the pace at which weapons develop is determined by the effectiveness of the procedures established to translate ideas into weapons. The prior acceptance and application of the thesis that superior arms favor victory, while essential, are insufficient unless the "superior arms" are accompanied by a military doctrine of strategic or tactical application which provides for full exploitation of the innovation. But even doctrine is inadequate without an organization to administer the tasks involved in selecting, testing, and evaluating "inventions." The history of weapons in the United States is filled with evidence on this point.

For want of an adequate administrative organization in the Ordnance Department, as shown earlier, Federal troops in the Civil War fought with inferior weapons even though better arms were available. There were at least two major factors contributing to the ineffectiveness of the methods used by the Ordnance Department to select weapons. The first factor was the apparent inability of the successive authorities to establish either a sound organization or effective administrative procedures to accomplish the desired task. The second, the pressure of an obvious need for standardization in opposition to the continual pace of technological development, is typified by the comment of Secretary of War Joel R. Poinsett in 1838 when he declared that Ordnance should "suffer a paralysis" rather than be "exposed to frequent changes and fluctuations." [49]

The Chief of Ordnance was officially responsible for the "patterns, forms, and dimensions" of all items purchased by Ordnance, but it had become customary for the chief to rely upon a board of officers "to adjust the details." Until 1839 appointments to this board had been made from all the various arms of the service, but from that date on the Ordnance Board was composed exclusively of officers from the Ordnance Department.[50] While this decision undoubtedly improved the technical qualifications of the board's membership, it also deprived the board of the point of view of the branches which used its services. Although there were serious dis-

48. Fuller, *Armament and History*, p. 163, and *The Reformation of War*, p. 110. See also Report of Surgeon General, *Annual Reports of the War Department*, 1920, Vol. *1*.

49. *Ordnance Reports, 3,* 356.

50. *Ibid., 3,* 225.

advantages in a board lacking the consumer's point of view, it might be argued that specialists, if working full time, could be expected to take a greater and more effective interest in improved weapons than any occasional and part-time board of constantly changing composition. Unfortunately, though, for the progress of weapons, as late as 1861 the Chief of Ordnance informed Secretary of War Cameron that while the establishment of a permanent board was desirable it was impossible, since all officers were engaged in the "pressing and indispensable duties of the Department." The Chief of Ordnance recommended that the plan to form a permanent board be "deferred to a future time." [51] It was decisions of this order which prevented Federal troops from fighting with the best available weapons and resulted in an unissued surplus of 1,195,572 obsolete muzzle-loading muskets at the end of the war.[52] Here was quantity, not quality.

The organization and functioning of the Ordnance Board, critical as it may have been, were by no means the only aspects of the administrative procedure which constituted the Ordnance Department's process for acquiring new weapons. Regardless of how well or how poorly any succession of ordnance boards may have performed their tasks, battle alone could be the final criterion of the value of a weapon, and this circumstance made necessary an adequate system for securing accurate reports from tactical units of the services in time of war and from military attachés and observers abroad during periods of peace at home. During the 10 or 20 years immediately preceding 1861 the Ordnance Department had sent occasional special observers to foreign nations to watch advances in weapons, which then appeared in the United States only "tardily after being matured abroad." Nevertheless, as late as 1853, even while recognizing that the limited experience of this nation in actual warfare made the department necessarily dependent upon the military services of other countries for improvements in weapons, the Chief of Ordnance regarded the idea of sending a technical mission abroad as advantageous but unnecessary in view of the high state of perfection of the arms issued by the department.[53]

If the procedure for reporting on foreign experience with weapons and exploiting foreign technological advances was haphazard and ineffective, almost exactly the same could be said about the system which the Ordnance Department had for securing re-

51. *Ibid., 3,* 226.
52. See above, p. 9 n. 17.
53. *Ordnance Reports, 2,* 290, 397, 531.

ports on the performance of weapons issued for use in combat. From the time of the Mexican War to 1861 there were few opportunities to secure operational reports. Thus little or nothing was done to establish a routing procedure for reporting back to the department the results of tactical experience with items in the field. In 1862 the Chief of Ordnance made an attempt to improve the situation. He asked his officers serving with troops in the field to keep daily notes of any "defects or deficiencies" in weapons and report them promptly with suggestions for "suitable remedies." This procedure, foreshadowing the system of rendering *Unsatisfactory Reports* evolved many years later, had all the weaknesses of the latter system in that it depended entirely upon the initiative of officers in the field and revealed trouble only after it had happened.[54]

Probably the real beginning of scientific accumulation of data for ordnance came after the war, in 1867, when orders went out to all batteries of artillery requiring an exact and detailed report of each shot fired. Units were instructed to record the history of each gun, the weight of projectiles, and the quality of powder used, as well as other similar information, on blank forms provided for the purpose by the Ordnance Department.[55] A few years later an imaginative and resourceful Chief of Ordnance, Brig. Gen. Stephen Vincent Benét, demonstrated the real utility of a systematic collection of statistics as a basis for decisions regarding development of weapons. Using the figures compiled by the Surgeon General on casualties during the Civil War and reinforcing them with similar statistics from the Franco-Prussian War, General Benét argued that the saber and bayonet were no longer important weapons. Presentation of these facts started the movement which reduced the saber to the status of ceremonial gear.[56]

To pursue this line of thought further would be to write the history of the Ordnance Department. It is quite unnecessary to do so, for the essential elements in the problem of the development of weapons can be studied in detail from the period already mentioned. The experience of the department demonstrated the importance of establishing a concept of requirements, the military characteristics of a weapon, before beginning development. Similarly, experience had shown the importance of differentiating

54. *Ibid., 3,* 438. The *Unsatisfactory Report* currently used in the USAF consists of an official form which units in the field are urged to use when reporting to higher headquarters on any unsatisfactory performance in equipment issued.

55. *Ibid., 3,* 313.

56. *Ibid., 3,* 101–2.

a good idea from the failure of that idea in a specific application. By the end of the Civil War there should have been no difficulty in recognizing the need for a service test to prove new weapons, for an adequate system to evaluate and report on performance in combat of new weapons, and for securing systematic reports on advances in foreign weapons. The problem of the organization and composition of an Ordnance Board, as well as the utility of statistical data on which such a board might base its decisions, could be studied in great detail before the turn of the century. In short, almost all of the problems which were to prove so vexing in the development of aerial weapons crowded the pages of Ordnance history.

The records of both the War and Navy departments were full of lessons of positive value to those responsible for the development of weapons in the years to come. Unfortunately, many of these lessons were buried in cluttered archives, virtually inaccessible to the officials who best could profit from them. Trained historians can sometimes bring the lessons to light but often too late to be of use. For example, James Phinney Baxter's analysis of the problem of developing weapons, which appeared in his naval classic, *The Introduction of the Ironclad Warship*, was not published until 1933, rather late to be of value to those charged with perfecting the aerial weapon. Nevertheless, it is perhaps significant that the substantial lessons to be garnered from the experience of the Ordnance Department were available, for the most part in published form, before the Wright brothers flew their first airplane.

The chapters that follow will deal with the problem of the aerial weapon from the period when flight first became a practical reality to the end of World War I. An attempt will be made to determine whether or not military leaders in the United States actively accepted and put into practice the thesis that superior weapons favor victory. An attempt will also be made to determine whether or not military doctrine was modified to meet the expanding potential of the new weapon and whether or not the proponents of the new weapon established an effective procedure for developing the innovation to justify modifications of doctrine. The evidence indicates that armies, war offices, and governments at the outbreak of World War I lacked effective systems for integrating the advances of science with the military machine. Anyone who seeks to evaluate the incorporation of the aerial weapon into the military establishment must recognize the problem as falling within this historical context.

*PART TWO: The Problem of the Air Weapon, 1907–18*

# Chapter II. The Air Weapon in the United States, 1907–18

FOR CENTURIES, from Da Vinci to Tennyson, visionaries have dreamed of the power that aircraft would some day unleash. Yet, when practical flight at last became a reality in the United States, the army was as slow in recognizing the implications of the air weapon as Congress was in appropriating funds for its development.

When Secretary of War William Howard Taft presented his annual report for 1904 to the President, the United States Army, a total force of 60,000 officers and men, was in a period of drastic transition. The Spanish-American War had come and gone, interrupting a lethargy of more than 30 years punctuated only by Indian wars. Elihu Root's reform of the General Staff, one of the many army heritages of the war, had already begun to take effect. The impressive innovations of the year were the Army War College and the new Springfield rifle. The secretary singled out the Chief of the Signal Corps for special commendation in his report. The Corps had shown "foresight and energy," he believed, in executing its projects during the year. Despite these words of praise, the secretary did not even mention airplanes or aeronautics although the Signal Corps was traditionally the military sponsor of aerial undertakings and the Wright brothers had already made more than 100 successful flights.[1]

The War Department had not always neglected aeronautics. Both the Union and Confederate armies used captive balloons for observation, and a captive balloon detachment saw service in the Spanish-American War. Unfortunately, the balloon crew at Santiago was untrained and attracted enemy fire by approaching too near the front lines; this aroused an unfriendly feeling toward aeronautics on the part of the ground troops. This early lesson

---

1. Report of Secretary of War, *Annual Reports of the War Department*, 1904, *1*, 22. See also A. Sweetser, *The American Air Service* (New York, D. Appleton, 1919), p. 8.

in the importance of air-ground relations was probably lost when the balloon detachment was inactivated at the end of the war.[2]

The War Department did not confine its interest in aeronautics to lighter-than-air equipment; by 1898 the Board of Ordnance and Fortification had assisted the famous Smithsonian scientist, S. P. Langley, with two separate grants of $25,000 to carry on his promising airplane projects. But the failure of Langley's trials on the Potomac in 1903 excited such a wave of editorial ridicule that the War Department, ever conscious of the need for Congressional appropriations, came to mistrust all inventors of heavier-than-air devices.[3]

The Wright brothers, for example, first approached the War Department in January 1905 after receiving a visit from a British officer who represented the Royal Aircraft Factory at Aldershot. Before starting negotiations with the British government, the Wrights determined to offer their airplane to their own government. Choosing a circuitous route, they wrote to a local congressman. When their proposition finally reached the appropriate agency, the Board of Ordnance and Fortification, that body regarded it with the suspicion customarily accorded suggestions proffered the War Department through political channels.[4] Although the Wrights had already flown as far as three miles at speeds up to 35 mph, the board brushed off their inquiry as if scotching an appeal for funds; yet, ironically enough, it offered to receive further representations as soon as the aircraft had been perfected "to the stage of practical operation."

Further attempts by the Wright brothers to demonstrate their airplane met with official apathy. In October 1905 the brothers again learned that the board did not care to formulate any requirement for an airplane "until a machine is produced which by actual operation is shown to be able to produce horizontal flight." The narrow attitude of the board prevailed until both the President

2. F. S. Haydon, *Aeronautics in the Union and Confederate Armies* (Baltimore, The Johns Hopkins Press, 1941); and Maj. F. P. Lahm, "History and Development of the Air Service," 13 Feb. 1920, National Archives, Bureau of Aircraft Production Miscellaneous Histories, Box 1; hereafter cited as NA, BAP Misc. Hist. For an explanation of archival materials cited in this study, the reader is referred to the footnote guide and glossary found at the beginning of the volume as well as to the bibliographical note at the end. See also AAF Historical Study No. 25, "Organization of Military Aeronautics; 1907–1935," pp. 1–5, Dec. 1944, AAF Archive.

3. Sweetser, *The American Air Service*, p. 5; and "History of the U. S. Army Air Service," 1 Oct. 1920, NA, BAP Misc. Hist. Box 1.

4. For an appreciation of the War Department's point of view on the problem of "inventions" forwarded by congressmen, consult any War Department organization's 400.111 file.

and the Secretary of War took a direct interest in the Wrights' cause.[5] In 1907, after sending Maj. G. O. Squier to Europe to study the progress of aviation there and Lt. G. E. Selfridge to Nova Scotia to study Alexander Graham Bell's aircraft experiments, the War Department established an Aviation Division in the Office of the Chief Signal Officer. One of the first acts of the new organization was to request bids for an aircraft and an airship.[6] Of the 24 bids received, only two led to contracts. Of these, only the Wright brothers produced a successful aircraft. In February 1908 they signed a contract to meet comparatively severe specifications—an airplane with a high speed of 40 mph, a range of 125 miles, and a useful load of 350 pounds. When Congress failed to appropriate the $200,000 asked by the Signal Corps for aviation needs in 1907, the Board of Ordnance and Fortification, as if to atone for earlier neglects, provided funds to cover the aircraft contracts. Flight trials held late in 1908 at Fort Myer, Virginia, showed the Wrights' machine was actually capable of better performance than that stipulated.[7] After many vicissitudes the army had a new weapon. The next 30-odd years were to be spent searching for ways to exploit that weapon more effectively.

From the date of the Wrights' first contract to the outbreak of war in Europe seven years elapsed. During that period the Signal Corps had to build an organization to deal with the new device and, at the same time, to develop and evaluate its potential role in the military establishment. The flying machine in 1908 seemed full of promise. The Secretary of War, apostrophizing the age of flight as "almost at hand," predicted that the airplane would some day "profoundly affect modern warfare." To implement this faith the secretary asked Congress for $500,000 to help the Signal Corps develop military aviation "in a manner commensurate with its intrinsic importance."[8] Despite the secretary's appeal, during 1909 and 1910 Congress failed to provide funds for aviation.[9]

In 1908 the aviation force of the Signal Corps comprised three officers and 10 enlisted men operating one airplane and three balloons. If this complement appears small, it must be remembered

5. Sweetser, *The American Air Service*, p. 8; and F. C. Kelly, *The Wright Brothers* (New York, Harcourt, Brace, 1943), pp. 164–5.

6. "History of the U. S. Army Air Service," n. 3, above.

7. Sweetser, *The American Air Service*, p. 8; and G. O. Squier, "Present Status of Military Aeronautics," Smithsonian *Report* (1908). See also *Annual Reports of the War Department*, 1908, *1*, 45.

8. *Annual Reports of the War Department*, 1908, *1*, 45.

9. Maj. G. R. Perera, "A Legislative History of Aviation in the United States and Abroad," March 1941, AAF Archive M1154–1.

that at the time the Signal Corps as a whole consisted of no more than 46 officers and some 1,200 men. Moreover, if Congress appeared slow in granting funds, it should be remembered that earlier disasters, such as those of the Langley machine and the crash at Fort Myer which killed Lieutenant Selfridge during the army's acceptance trials of the Wrights' airplane, were reason enough for appropriating money with caution.[10] The lack of funds for aircraft was not entirely the result of Congressional suspicion, however. Elihu Root's reforms were still in process throughout this period, and the War Department gave first priority to appropriations for modernizing such "absolutely necessary" materiel as field guns and ammunition. Despite these handicaps, by 1913 the Signal Corps had 22 aircraft on hand or on order and 14 pilots trained to fly them.[11]

In view of the extremely limited number of aircraft available and the totally inadequate number of men trained to use them, the Signal Corps made a series of significant advances before 1914. Even the restricted operations which it was possible to undertake with the facilities on hand led to the conclusion that commercially procurable machines were inadequate for military purposes and that it would therefore be necessary to draw up specifications for tactical aircraft.[12] To ascertain the requirements for specifications defining the desirable standards of performance for tactical aircraft, the Chief Signal Officer devised a program requiring the establishment of several aviation centers. According to the proposed program, these centers would be used not only to train pilots and mechanics but also to study the design of aircraft, to perform tests, and to consider "all other matters tending to improve the military aviation services."[13]

The program provided an organizational basis upon which the development of aircraft as an aerial weapon could proceed systematically. Limited appropriations of course prevented the program from being executed in full, yet some important experiments were actually carried out. Flight tests proved the feasibility of firing upon ground targets with an aircraft machine gun. At about the same time a bombsight was repeatedly improved until it performed successfully.[14] A detachment of aviators from the Signal

10. AAF Historical Study No. 39, "Legislation Relating to the Air Corps Personnel and Training Programs, 1907–1939," Dec. 1945, AAF Archive.
11. AAF Historical Study No. 25, p. 5. See also *Aeronautics in the Army,* Hearings before the House Military Affairs Committee, 63 Cong. 1 Sess., 1913.
12. "History of the U. S. Army Air Service."
13. AAF Historical Study No. 39, p. 120.
14. "History of the U. S. Army Air Service."

Corps continued bomb-dropping experiments in 1914 until the need for every available aircraft to train the increasing numbers of pilots forced experimental activity into the background.[15] Thus, by the time war came to Europe in 1914, the airplane had been demonstrated in the United States as a valuable weapon. Flight trials and limited service tests had already marked out the directions in which the potentialities of aircraft might evolve. The degree to which the War Department exploited these several potentialities during the years to follow rested almost entirely upon the ability of the department to organize its establishment to embrace a revolutionary concept.

If, then, on the eve of war the United States could take pride in both the recognition and adoption of a novel weapon, the other Powers were even more positive in their appreciation. The following table of aeronautical appropriations during the 1913 fiscal year provides a useful yardstick.[16]

| | |
|---|---|
| France | $7,400,000 |
| Germany | 5,000,000 |
| Russia | 5,000,000 |
| England | 3,000,000 |
| Italy | 2,100,000 |
| Mexico | 400,000 |
| United States | 125,000 |

Mexico, though scarcely a Power, had appropriated more in one year than the United States had in six years. The total expenditure of the War Department for aircraft from 1908 to 1913 amounted to a mere $250,000. A comparison in 1914 of available aircraft and trained pilots proves equally revealing.[17]

| | Aircraft Available | Pilots Trained |
|---|---|---|
| France | 260 | 171 |
| Russia | 100 | 28 |
| Germany | 46 | 52 |
| United Kingdom | 29 | 88 |
| Italy | 26 | 39 |
| Japan | 14 | 8 |
| United States | 6 | 14 |

15. Report of Chief Signal Officer, App. I, *Annual Reports of the War Department,* 1914, *1,* 522.

16. Report of Secretary of War, *Annual Reports of the War Department,* 1913, *1,* 25–6. Figures from other sources vary slightly but not significantly.

17. Sweetser, *The American Air Service,* p. 16. Perera, "A Legislative History of Aviation," offers somewhat different figures, but the ratios between nations remain substantially the same.

Moreover, in terms of experimental activity Europe showed no want of interest. Before the end of 1914 Rudolf Boehm in Germany had flown a Rumpler for 24 hours nonstop, Igor Sikorsky in Russia had completed a four-engine aircraft (the first so built), and in England Short Brothers, Ltd., mounted an experimental two-pounder naval gun in an airplane even before war came.[18] Of still greater importance, the Royal Flying Corps, established with a military and naval wing in 1912, formed an Experimental Branch in 1913. By the end of another year more than a dozen English manufacturers were turning out aircraft at the rate of 100 a year.[19]

Thus by 1914 the British had already established an organization specifically charged with experimentation. In the United States, where no such separate organization existed, experimental projects were subject to a lower priority than training. In addition, where industry had shown little enthusiasm for aviation in the United States, Britain's aircraft industry, while young, was already established and expanding rapidly when the war came. Historically, most new ideas for weapons have encountered delays in application. In regard to the air weapon, the evidence suggests that the delays it met in the United States were comparatively more serious than those encountered in foreign countries.[20]

Congress responded to the stimulus of troubled Europe and acted by means of remedial legislation to reduce the lag between aircraft advances in the United States and those abroad. An act of July 1914 sanctioned the Aviation Section of the Signal Corps, hitherto an administrative creation capable of being abolished on the order of the branch chief. The act created a section consisting of 60 officers and 260 enlisted men.[21] It also provided for special aviation ratings and flying pay which attracted increasing numbers of men into the service, but of more importance from the point of view of materiel, the permanent nature of the legislation induced manufacturers to run the risk of building aircraft for military purposes.

Substantial as the benefits of the new legislation were, they did not constitute a revolutionary step toward incorporation of the

18. C. G. Grey, *The History of Combat Airplanes* (Northfield, Vt., Norwich University, 1941), pp. 2, 17.

19. H. A. St. G. Saunders, *Per Ardua* (London, Oxford University Press, 1945), pp. 19–21, 212.

20. AAF Historical Study No. 50, "Materiel Research and Development in the Army Air Arm; 1914–1945," p. 12, Nov. 1946, AAF Archive.

21. AAF Historical Study No. 25, p. 18; 38 Stat. 514, Public No. 143, 18 July 1914.

air weapon in the military establishment. That Congressional action would effect decisive changes in the status of the air weapon was hardly to be expected. Even the Signal chief, who was primarily responsible for the development of the new weapon, doubted the alleged potentialities of aviation. He admitted the influence which aircraft would exercise on tactics as a direct result of their use as observers both for adjustment of artillery-fire and for general reconnaissance; at the same time he questioned the offensive value of aviation. In testimony before the House Military Affairs Committee in December 1914, he said bluntly, "As a fighting machine the airplane has not justified its existence." [22]

The Chief Signal Officer was perpetuating a thought pattern that the experience of the Ordnance Department during the 19th century had shown to be only too prevalent. His testimony exposed his failure to distinguish between the concept of the aerial weapon and its contemporary application. The probable span between existing technological development and ultimate development escaped him. The Chief Signal Officer's attitude was scarcely the reflection of a careless moment: he repeated the stand in submitting his annual report to the Secretary of War; and apparently befriending the cause of aviation in appealing for more airplanes, he confirmed his earlier views by going out of his way to declare that European experience had shown offensive use of aircraft to be so unwarrantably expensive as altogether to "discredit general attack by air." The verdict on offensive aviation, the chief believed, should rest at "not proven."

The opinion of the Chief Signal Officer was of exceptional importance, because as the administrative officer most intimately associated with the innovation his judgment and recommendation must surely influence those at higher echelons charged with broad decisions on strategy and materiel.

The tactical and strategic utility of the airplane as a weapon could be demonstrated only by an interminable succession of service tests following each technological advance in the whole field of aeronautics. Such tests and advances in engineering would absorb endless funds, but since the Chief Signal Officer had little faith in the potentialities of aircraft as offensive weapons there was little likelihood that he would push Congress to grant appropriations for aviation. Consequently the Signal Corps left the concept of the

22. Quoted in Sweetser, *The American Air Service*, p. 26. See also Brig. Gen. G. P. Scriven, *The Service of Information* (Washington, 1915), Office, Chief Signal Officer, Circular No. 8.

air weapon only partially explored. The Chief of the Signal Corps may have believed that he was just being appropriately cautious when he said, "Little of importance has been proved" by the appearance of an airplane able to drop "bombs and other missiles," [23] but his words apparently had far-reaching effect. Representative McKellar, speaking before the House Military Affairs Committee in December 1914, remarked that inasmuch as aircraft had "proved worthless to a very large extent" the United States could be considered "gainers by not having spent so much." [24]

Thus, for want of a broad appreciation of the short span between existing airplanes and their evolutionary possibilities, the future of the air weapon lay in jeopardy. Representative McKellar's attitude may not have been typical, but Congressional appropriations in the years leading up to 1914 would indicate that it was. The congressman could not have perceived that a dynamic, aggressive aircraft industry is as vital to the air weapon at the outbreak of war as is a fleet-in-being to the naval arm.

When one considers the lack of Congressional enthusiasm and the grudging support rendered by the leaders of the Signal Corps, it is surprising that those actually charged with developing the novel weapon moved as far as they did. The problem of aircraft development was tremendously complicated by the prevailing uncertainty regarding tactical objectives; they were never clearly defined before the United States entered the European conflict. Unlike most weapons, the airplane was capable of performing several tactical functions. Each of these required a specialized line of technical development.

Acceptance of the new weapon by higher echelons of command depended in great measure upon test demonstrations and upon actual performance in combat. The emphasis placed by officials of the Signal Corps upon any one of the several specialized tactical types determined which type would secure an opportunity for trial in combat. Only by such a test could any new type of weapon prove its utility and win for itself financial support as well as the confidence of commanders in the field. Since the tactical function of observation required far less specialized equipment than did bomber and fighter aircraft, it was perhaps inevitable that the value of observation aircraft should win recognition before the technologically more complex fighter and bomber. But this very

23. Report of Chief Signal Officer, *Annual Reports of the War Department*, 1914, *1*, 505–8.
24. Quoted in Sweetser, *The American Air Service*, p. 27.

fact only serves to point up the tremendous responsibility which rested in the hands of those developing materiel in the new aviation center. The more successful they were in defining and refining exact statements of the tactical objectives and performance characteristics they wanted in fighters and bombers, the greater was the likelihood of securing desirable aircraft from industry. For how could industry design specialized types of airplanes before the Army was able to tell what would be expected of these airplanes in combat? Increasing success in obtaining effective tactical aircraft types reduced to practice would mean greater likelihood of obtaining convincing tests in combat, which in turn could lead to increased financial support as well as the cooperation of field commanders.

In short, advocates of the air weapon within the Signal Corps could expect to see the fullest exploitation of the potentialities of aircraft only insofar as they succeeded in setting the pace, that is, in defining objectives for the aircraft industry. The organization established to determine what those objectives should be bore a weighty responsibility for the destiny of air power in the United States.

In 1915 primary responsibility for the evolution of military aircraft in the United States rested with a handful of men stationed at the North Island aviation center in San Diego Bay. The center had two departments, one for training, the other for experiment and repair. The staff of the latter department consisted of one officer in charge, one civilian aeronautical engineer, one civilian mechanical engineer, and five civilian mechanics. Their duties at the base in addition to "the study of new types" included overhauling, repairing, and rebuilding the training aircraft as well as maintaining equipment for ground servicing.[25] Small as this organization was, it provided a nucleus from which to expand.

While it was undoubtedly true that maintenance functions absorbed a large portion of their time, the members of the staff did not entirely neglect the development of new equipment. During the preceding year, in fact, the Aviation Section had announced a competition to be held at the San Diego center to secure data "to enable the school to decide on a standard machine." The phrase "a standard machine" is revealing, for it indicates that in 1914 the concept of markedly differentiated tactical types had not yet taken root. Although it appears that there had been little official

25. Report of Chief Signal Officer, *Annual Reports of the War Department*, 1915, *1*, 744–5.

speculation regarding the doctrine of air power, nevertheless, during 1914 the first specification for a military aircraft appeared in the United States in connection with the competition. The Signal Corps drafted requirements for a "reconnoissance [*sic*] aeroplane," a two-place biplane capable of lifting a useful load of 450 pounds at a high speed of 70 mph.

The appearance of specifications for a military aircraft was certainly a significant move toward recognition of the tactical value of the air weapon, but equally meaningful was the procedure established to select the new weapon: a competition in which points were awarded for superior performance as demonstrated by actual flight of the airplanes submitted by rival manufacturers. Speed and rate of climb were to be measured by objective test, maneuverability by performance of prescribed evolutions, and field of vision by test with a military observer; construction and standards of workmanship were to be determined in a "practical examination" by the evaluating board.[26] Here, for all its obvious shortcomings, began an administrative mechanism for selecting superior weapons. Of particular interest was the recognition given to tactical suitability, especially as demonstrated by testing the field of vision for observation purposes. This, after all, was the mission of the aircraft and properly deserved full consideration along with purely engineering factors. Regrettably, in the years to follow, evaluating boards sometimes lost sight of the primary importance of tactical suitability.

The competition of 1914 brought in 12 different bids, but the lack of a reliable aircraft engine fixed an upper limit on the enhanced performance to be expected from these new designs for airframes. So critical, indeed, did the engine problem become that during 1915 the Signal Corps determined to stage an engine competition similar to the aircraft contest already held. During these competitions and subsequent flight trials the several manufacturers supplying aircraft began the practice of maintaining representatives at the aviation center "to keep in touch with the needs of the new section." [27] Thus, by the end of 1916, the Signal Corps had established a method for selecting superior weapons and a close working relationship with the manufacturers supplying them. The number of aircraft actually procured and the number of pilots

26. Report of Chief Signal Officer, Appendix I, *Annual Reports of the War Department*, 1914, *1*, 517–21.

27. Report of Chief Signal Officer, *Annual Reports of the War Department*, 1914, *1*, 9, and 1915, *1*, 746.

available to fly them remained woefully small, restricted as they were by insufficient appropriations. Within the limits imposed it might be argued that the development of the air weapon was progressing normally. In Europe, however, the pace of development was no longer normal.

Somewhat later in 1916 the extreme importance of the airplane as demonstrated in the European war began to win recognition in the United States. The National Defense Act, passed in June, strengthened the Aviation Section by authorizing increases in personnel and providing for special flying pay, but even with this encouragement only 43 officers received pilot training during the year. Nevertheless, the substantial increase in money available, some $500,000 in deficiency funds over and above the annual appropriation of $300,000, marked a turning point second in importance only to the rapidly evolving doctrine of air warfare.[28]

Eighteen months after the outbreak of war in Europe the Chief Signal Officer was ready to retire somewhat from his former position and to admit the need for three different types of military aircraft: a reconnaissance and artillery fire-control type, a combat type, and a pursuit type. To be sure, the differentiating characteristics of these three types were vague, but the admission of their separate existence was a step toward fuller exploitation of the air weapon. Basing his judgment on "conclusions reached from experience abroad," the Chief Signal Officer continued to feel that the "most important work of aircraft" lay in performing missions of observation. Nonetheless, he now considered the utility of aircraft for liaison purposes and for defense against other aircraft to be "obvious." Accordingly, with this new recognition of the multiple role of aviation, the chief proposed to establish squadrons of 12 aircraft, in which eight were to be observation type, two were to be "rapid flying machines for chase or transport," and two were to be of a bomb-carrying or offensive type.[29]

In the spring of 1916, Lt. Col. G. O. Squier, an officer of the Signal Corps with a reputation as an engineer extending beyond army circles, assumed command of the expanding Aviation Section. As an army engineer with university training, Colonel Squier tried to apply scientific method in the procurement of designs for

28. Report of Chief Signal Officer, *Annual Reports of the War Department*, 1917, *1*, 839, and 39 Stat. 174, Public No. 85, Sect. 13, 3 June 1916. See also AAF Historical Study No. 39.

29. Report of Chief Signal Officer, *Annual Reports of the War Department*, 1915, *1*, 742, 747. See also *Military Aviation* (Washington, 1916), prepared as War Department Document No. 515 by the War College Division of the General Staff.

aircraft. He wished to incorporate the "lessons gained by experience," that is, actual operation under combat conditions. The colonel's objective was undoubtedly sound, but the Signal Corps lacked facts from which to proceed. The punitive expedition across the Mexican border was the only tactical operation in which the army's aircraft had participated. One squadron of eight aircraft operated for about six weeks until all the available equipment was smashed.[30] This unhappy episode stimulated technical improvement and emphasized the importance of facilities for maintenance of aircraft in the field, but the limited number of aircraft involved, the lack of aerial opposition, and the extremely short period of operation all militated against the hope of learning many significant tactical lessons from the Mexican expedition.

Denied a body of combat experience in Mexico, officials of the Signal Corps sought every possible lesson from the European war. Such lessons proved extremely difficult to acquire because of the barriers of secrecy with which the warring Powers surrounding their aircraft. Beginning in the first few months of the war, for reasons of security, the Powers imposed restrictions which amounted to a virtual blackout of detailed tactical and technical information emanating from Europe, information which was essential to the Aviation Section in the United States. Alarmed by the situation, a Congressional committee demanded a statement as to the relative position of the United States in aviation as compared with the countries of Europe. A representative of the Signal Corps admitted that the Aviation Section was keeping abreast of Europe only insofar as it was possible to say the section was keeping abreast of conditions of which it knew nothing. Secretary of War Newton D. Baker informed the House Military Affairs Committee in 1916 that the censorship in Europe was exceedingly strict on aviation matters. For a knowledge of trends in the development of aircraft the United States had to rely upon manufacturers in this country working on contracts from the warring nations. Embassies of the belligerents repeatedly rejected requests by the War Department for permission to send qualified aviators to the fighting fronts as observers.[31]

30. AAF Historical Study No. 25, pp. 26–7; and Report of Secretary of War, *Annual Reports of the War Department*, 1916, *1*, 40. See also Testimony of Maj. B. D. Foulois, *War Expenditures*, House Hearings, 66 Cong. 1 Sess., Serial 2, Pt. 6, 6 Aug. 1919, pp. 344–5.

31. Testimony of Maj. B. D. Foulois, 6 Aug. 1919, *War Expenditures*, House Hearings, 66 Cong. 1 Sess., Serial 2, Pt. 6, 6 Aug. 1919, p. 348; and Sweetser, *The American Air Service*, pp. 26, 35.

When the United States finally did declare war on Germany, there were exactly five aviation officers abroad: three were attending flying schools in France; one was serving as an assistant military attaché in London; and one, Maj. William Mitchell, had at last succeeded in securing permission from the French government to go to France as an observer. None of these officers had been abroad more than a few weeks when the United States entered the war, and none had had sufficient experience to report significantly on the status of operational aircraft in Europe.[32] The information received by the Aviation Section through military channels up to the time the United States entered the war was "so meagre as to be useless."[33]

The condition of the air weapon in the United States when war ultimately came was extremely critical. The organization charged with developing the weapon seemed inadequate for the task; there existed only the haziest notions regarding the doctrines of aerial warfare. In terms of equipment and numbers, the nation was several years behind development in Europe. The limitations imposed by the shortage of funds and authorized personnel had restricted growth to the point where only the most circumscribed range of operating experience was possible. While the aircraft and engine competitions providing for procurement on the basis of performance laid an important precedent, the process had not been formalized by regulation or legislative action for continued use and perfection. In April 1917 the Aviation Section consisted of 65 officers and 1,120 enlisted men. Flying activities took place at two airfields with 200-odd airplanes, mostly training types, "nearly all obsolescent."[34] Such was the character of the air weapon with which the United States entered the war in Europe. Moreover, the paucity of domestic operational experience and the absence of close liaison with "Europe's startling developments" in aviation left the United States without a basis from which to derive a doctrine of air warfare.[35] With no doctrine, or at best with a vaguely de-

32. *War Expenditures*, House Hearings, 66 Cong. 1 Sess., Serial 2, Pt. 31, 6 Aug. 1919, p. vii.

33. Col. E. S. Gorrell, "Early Activities of the Air Service, AEF," *ca.* 1919, National Archives, World War I Organization Records, Air Service Historical Records, Box 2; hereafter cited as NA, WWI Orgn. Records, A.S. Hist. Records.

34. B. Crowell, *America's Munitions; 1917–1918* (Washington, GPO, 1919), p. 240. Lahm, "History and Development of the Air Service," gives different figures, claiming only 55 aircraft were available 6 April 1917. Of these, he adds, 51 were obsolete, 4 were obsolescent. Crowell's figures are more reliable.

35. Report of Director of Military Aeronautics (DMA), *Annual Reports of the War Department*, 1918, *1*, 1381–2.

fined doctrine, the development of military aircraft was bound to suffer.

How these limited resources in organization, doctrine, and equipment were transformed during the years to follow will be described in subsequent chapters.

## Chapter III. Planning the Aerial Weapon

When President Wilson signed the measure officially declaring war on Germany in 1917, those who planned to use the new aerial weapon lacked a clearly defined doctrine of warfare. The initial step in solving the problem lay in securing aircraft superior to those employed by the enemy. The procurement of superior aircraft implied two prior assumptions: first, a knowledge of the mission of the new weapon; and second, a knowledge of the types of aircraft necessary to accomplish this mission. Before an air force could take shape, it was also essential to implement aerial doctrine by determining the composition of the air arm. If the mission of aviation was observation only, then the composition of the air arm would consist solely of observation aircraft. If, on the other hand, this mission included defense and offense, then obviously the composition of the air arm should include fighters and bombers. Once the functions to be performed were decided upon, it would then be necessary to determine the relative proportion of each type which available resources—men, money, materials, facilities and transportation—would make feasible. Then, finally, it would remain to determine the performance characteristics required to ensure individual types of aircraft superior to those sent out by the enemy. These decisions had yet to be made.

Eventually all these decisions were made simultaneously. Apparently they were made without a realization of the importance of determining doctrine before settling detailed questions regarding characteristics for each type. This situation may explain the troubles encountered in the aviation program as the war progressed. Never was the absence of an adequate organization for making decisions more acutely evident in the air arm than in April 1917.

Three days before the declaration of war the National Advisory Committee for Aeronautics (NACA), a research agency entirely separate from the Signal Corps, took the initiative in surveying the aircraft industry in the United States to ascertain its readiness to meet the demands shortly to be thrust upon it. A telegram which the NACA sent to every known manufacturer of airplanes laid bare the prevailing want of organization and policy. The message read in part: "Can you provide training reconnaissance airplanes? If so, state type . . ."[1] The implications of a policy which, for lack of guidance, left the determination of types to industry were soon apparent.

The Secretary of the Navy, Josephus Daniels, after negotiations with C. D. Walcott, chairman of the NACA executive committee, and with the concurrence of the Secretary of War, took the initiative in establishing a Joint Army-Navy Technical Board. The new board was to "standardize as far as possible" the designs and general specifications of aircraft to be procured by the services.[2] Standardization required agreement on types and designs. Agreement involved a prior decision regarding the composition of the air arm. The composition of the air arm represented doctrine reduced to practice. A statement by higher authority regarding the relative proportion of functional types in the proposed force would have constituted by implication a directive on doctrine. In the absence of any such directive, the Joint Army-Navy Technical Board set about determining the composition of the air arm on its own initiative. As a consequence, formulation of aerial doctrine fell by default to a technical board officially charged with making nothing but technical decisions. The board, consisting of six officers, three from each of the two services, reported directly to the two secretaries. Although in this position it was in an echelon close enough to the top councils of war to receive official pronouncements of doctrine, none seems to have been sent.

The only significant indication of the existence of any air policy at all appeared in the program of the General Staff for a million-man army hurriedly drafted just prior to the declaration of war. This program provided for the organization of 16 aviation squadrons, one for the headquarters of each army corps, but contained no analysis of the composition of the force so raised. The Joint Army-Navy Technical Board may have been influenced but was certainly not bound by this proposed organization when a state of war actually materialized. The board devoted the first few weeks

1. Draft copy of NACA annual report, 20 October 1917, Air Force Central Files, hereafter cited as AFCF, 334.8 NACA.

2. Secretary of Navy to Secretary of War, 27 April 1917, AFCF, 334.7 Army-Navy Joint Boards. See also Secretary of Navy to C. D. Walcott, NACA, 9 April 1917, and Walcott to Dir. Council of National Defense, 10 April 1917, NA, BAP Hist. Box 9, 334.7 Joint Army-Navy Technical Board.

of its existence to drafting a program to procure training aircraft, leaving the determination of types for tactical or combat use until further advice could be secured from abroad.[3]

The program of the General Staff in March 1917 had assigned to aviation a relatively insignificant part in the organization for offensive operations, but the arrival of French and British military missions completely revolutionized this point of view. For want of better instructions, the Joint Army-Navy Technical Board relied upon suggestions from foreign officers and upon "vivid imagination" in framing a program.[4] The effort of the foreign officers was, of course, directed toward enlarging all previous estimates many times over. But mere increases in numbers, important as they were in conditioning everyone concerned to the almost astronomical sums that must be involved, did not determine the proportion of observation, fighter, and bomber aircraft desired— if indeed all of these were desired.

The board, in trying to draft a program for an effective air arm without adequate information from which to proceed, continued to flounder for nearly a month. Then, unexpectedly, Premier Ribot of France cabled to the French ambassador in Washington and laid down a detailed program of the aircraft desired as a contribution of the United States to the war effort. In this program, as received by the Joint Army-Navy Technical Board, the premier proposed the formation of an air arm of 4,500 aircraft for the campaign of 1918 with a monthly rate for replacement and reinforcement of 2,000 aircraft and a commensurate production of engines. The cable also urged the training of pilots and mechanics in numbers sufficient to man the force created. Ribot's message provided just what the board needed. Here at last was an arbitrary quantitative basis on which to formulate the whole program of development. The cable established a definite target for purposes of planning. A Signal Corps officer, Maj. B. D. Foulois, who was one of the army representatives on the Joint Army-Navy Technical Board, drew up a detailed program for production on the

3. "The Aircraft Production Board," *Proceedings of the Acad. of Pol. Sci., 7,* No. 4 (Feb. 1918). See also A. Sweetser, *The American Air Service* (New York, D. Appleton, 1919), pp. 44, 52, 60; and W. F. Willoughby, *Government Organization in War Time and After* (New York, D. Appleton, 1919), chap. xiv, for brief, general descriptive accounts of the organization.

4. Testimony of Maj. B. D. Foulois, *War Expenditures,* House Hearings, 66 Cong. 1 Sess., Serial 2, Pt. 6, 6 Aug. 1919. See also Report of Director of Military Aeronautics, *Annual Reports of the War Department,* 1918, *1,* 1381–2, and Memo, M. J. Grogan, 26 April 1919, NA, BAP Hist. Box 9, 334.7 Joint Army-Navy Technical Board.

basis of the premier's message. Toward the end of June 1917 Maj. Gen. Tasker H. Bliss, the Acting Chief of Staff, formally approved the program drafted to comply with the French request. In this manner the cable became the foundation of the nation's program for aviation.[5]

Inasmuch as Ribot's cable became the basis for the whole aviation program in the United States, a careful study of its origins is in order. As soon as it became evident that the United States was about to enter the war, the General Staff of the French army drew up a staff study on the composition of the air forces to be contributed by the new ally. This study, approved by the commander-in-chief of the French forces, set forth three distinct categories of air weapons. "First and foremost" was a group to be used in tracking down submarines; "of secondary necessity" was a group to consist of pursuits, bombers, and transports; and of third and last priority was an army group for service with large units of the expeditionary force. In this last group would be included all aircraft for observation, liaison, and artillery fire-control. "Aviation for pursuit and bombing operations," the French study pointed out, could go on increasing in size until the end of the war. This assumption, the study indicated, was "unquestionable." But for that category serving with the armies a fixed upper limit was established, equating the number of aircraft desired directly with the organizations served. This third category was to consist entirely of aircraft for observation, one squadron for every army corps and one for every regiment of artillery. Each army was to have two additional squadrons for reconnaissance. Apart from a nominal reserve, these allocations fixed the total number of squadrons required for observation or army-cooperation.[6]

The study by the French General Staff represented French aerial doctrine as formulated under fire during 1916 and early 1917, but it was not the quantitative basis for Ribot's cable. At about the same time that this study was prepared, the commander-in-chief of the French Armies of the Northeast drafted a plan for aerial participation by the United States. The memorandum sug-

5. Translation of cablegram, Premier Ribot of France to French Ambassador in Washington, 23 May 1917; Memo, Asst. Chief of Staff to Chief of Staff, 23 June 1917; and Memo, Act. Chief of Staff to Adjutant General, June 1917; NA, BAP Hist. Box 6, 311.2 Ribot Cable.

6. "Contribution to aviation to be demanded of the United States," translation of extract from French Army General Staff Study, April 1917, NA, BAP Hist. Box 6, 311.2 Ribot Cable.

gested that the ideal contribution would consist of 30 pursuit groups and 30 bomber groups, each group to comprise six squadrons of 12 aircraft per squadron, a total of 4,320 aircraft. With a margin for reserve, this figure could be rounded off at 4,500 aircraft, the basic figure stipulated in Ribot's cable.[7] In all probability the French premier had seized upon the convenient memorandum of the commander in the Northeast and sent it to Washington as a definite and tangible point of departure for the planners.

The cable, as received, was expressed entirely in quantitative terms: nothing in the message gave any indication of the composition of the 4,500 aircraft in terms of the relative proportion of fighters, bombers, or pursuits to be constructed. Both the study of the French General Staff and the memorandum from the commander of the Northeast Army had emphasized the primary importance of pursuit and bombardment as opposed to observation, liaison, and artillery-fire adjustment for the ground forces. The two military papers expressed, by implication but nonetheless clearly, a doctrine of air power. Since Ribot's cable adopted only the quantitative considerations from one of the military studies and ignored the implicit doctrine, the message failed to have the effect that the authors of the military studies had intended.

In the absence of a doctrinal precept from France, those who planned the program of aviation in the United States used the 4,500 aircraft and the factor of 2,000 per month for replacement as a purely quantitative guide. They resolved the question of composition according to their own ill-defined ideas of doctrine, which tended to attach greater importance to observation and proportionally less importance to bombing. Ironically, the evidence now available indicates that the number 4,500 mentioned in the French cable was intended to apply only to strategic aviation. This great separate air force was to operate independently of the units assigned to the field armies.[8] But the Joint Army-Navy Technical Board drew up a program using the basic 4,500 figure and the figure for subsequent monthly additions as if they included both strategic aviation and the tactical forces to be assigned to the ground forces. The board's understandable misinterpretation was to have far-reaching effects, for the initial program drafted on the

7. Memo, Commander-in-chief, French Armies of Northeast to Minister of War, 6 May 1917, quoted in "Air Service Programs, 1917," study for Hist. A.S. AEF, NA, WWI Orgn. Records, A.S. Hist. Records Box 300.

8. "Air Service Programs, 1917," study for Hist. A.S. AEF, pp. 5–6, NA, WWI Orgn. Records, A.S. Hist. Records Box 300.

basis of the French cable determined in a large measure the character of the air force ultimately sent to France by the United States.

Not until after the war ended did the historical accident which had occurred come to light. A comparison of the text of Ribot's message as received by the Joint Army-Navy Technical Board with the text of the message as received by the French Embassy in Washington revealed discrepancies. An additional phrase, apparently inserted by an overzealous member of the French military mission who had carried the cable from the embassy to the War College, injected a time element in the message. The alteration added the phrase "during the campaign of 1918," which imposed an obligation of early delivery. Another phrase, following the part concerning quantities, added, "which would allow the Allies to win the supremacy of the air." [9]

In all probability neither of these two alterations had any significant influence upon the program, but the discovery that the cable had been tampered with led an officer of the Air Service to investigate the background of the message more narrowly. In his searchings he learned that in July 1917 Ambassador W. G. Sharp had cabled the Secretary of State in Washington to say that Ribot's message as drafted in France also mentioned that the 4,500 aircraft should consist of "half bombers and half fighters." The "other necessary types," presumably aircraft for observation and army-cooperation, should be reckoned in addition to this total. [10] Whether this vitally significant portion of the message was omitted by design or by accident and whether the deletion took place in France or in the United States are unknown. In any event, the omission of the phrase "half bombers and half fighters" was crucial. Lacking these five words, the French cable was devoid of doctrine. Since there was no specific precept from France on aerial doc-

9. Testimony of Col. E. S. Gorrell, *War Expenditures,* House Hearings, 66 Cong. 1 Sess., Serial 2, Pt. 31, 24 Oct. 1919, p. viii. See also E. S. Gorrell, "What, No Airplanes?" *Journal of Air Law and Commerce* (Jan. 1941).

10. Memo, Capt. J. L. Ingoldsby, Chief, Hist. Sect. A.S., 4 June 1919, NA, BAP Hist. Box 6, 311.2 Ribot Cable. The full text of the Ribot cable as received by the Joint Army-Navy Technical Board is quoted in Sweetser, *The American Air Service,* p. 66; he was apparently unaware of the important alterations made in the message. Similarly, Gen. J. J. Pershing quotes in full the text as received in *My Experiences in the World War* (2 vols.; New York, Frederick A. Stokes, 1931), *1,* 28. Although the alterations of the text were discovered in 1919, General Pershing appears not to have known of them. Gen. H. H. Arnold's volume of memoirs *Global Mission* (New York, Harper, 1950), p. 50 perpetuates the error. General Arnold credits General Mitchell with having inspired the French cable, but the evidence is not conclusive.

trine, the decisions which shaped policy in the United States were reached from a synthesis of the limited and unofficial advices received from Europe before the cable arrived.

The Joint Army-Navy Technical Board sent its aviation program to the secretaries of War and Navy for approval on 29 May 1917, five days after the arrival of the message from France. The proposed plan stipulated the creation of a combat force of 3,000 aircraft for reconnaissance or observation, 5,000 fighters, and 1,000 bombers, with a reserve of 1,000, 1,667, and 333 in the three categories respectively. This force amounted to a total of 12,000 aircraft, the number contemplated by Premier Ribot as essential for the first six months of 1918 in addition to the original force of 4,500.[11] The sheer size of this force, larger than any responsible officer had even dreamed of before the war, made an adverse impression on the General Staff. For one thing, the proposed program would require a staggering appropriation. Moreover, officials on the General Staff feared lest the vast amount of material required would affect "everything else in the United States." [12] There was some justice in their alarm as the subsequent industrial mobilization was to demonstrate, but the ensuing delay in reaching a decision threatened to wreck the program. General G. O. Squier, who was responsible for the Aviation Section, took the problem directly to the Secretary of War over the heads of the General Staff. The secretary presented the program to Congress where 640 million dollars, one of the largest single appropriations made up to that time, was hurriedly voted.

The speed with which the aviation bill passed and the size of the sum appropriated were compelling evidence of the popular faith in aviation as a weapon for winning the war. Nevertheless, an enthusiastic public while important in winning financial support, did not differentiate between the functions of the several types of aircraft. Congressional comment had publicized the belief that one

11. Joint Army-Navy Technical Board to Secretary of War, 29 May 1917. For an indication of the program drafted before receipt of the Ribot cable, see JANTB to Secretary of War, 23 May 1917, NA, BAP Misc. Hist. Box 1, 321.9 A.S., Training. The Ribot cable is officially declared to be the basis of the United States program in "Final Report of the Chief of the Air Service, AEF," *Air Service Information Circular, 2,* No. 180 (15 Feb. 1921), 23.

12. Testimony of Maj. B. D. Foulois, *War Expenditures,* House Hearings, 66 Cong. 1 Sess., Serial 1, Pt. 6, 6 August 1919, pp. 373–4. Gen. P. C. March, writing some 15 years later in *The Nation at War* (Garden City, Doubleday, Doran, 1932), p. 201, calls the French request "ridiculous" and "preposterous." Significantly, however, he did little to alter the objectives of the program when he became Chief of Staff.

airplane was worth a regiment of cavalry.[13] To Congress, and presumably to the public at large, this contention, right or wrong, lumped all aviation into one category. In actual point of fact, the composition by types of the air force to be constructed with the huge appropriation would determine the relative value of the air arm in warfare. The success or failure of the aerial weapon in the war depended in a large measure upon the decision of the Joint Army-Navy Technical Board.

Because the significant French military studies which emphasized the importance of bombers and fighters over observation were unknown to them, the board members had drafted the program on a ratio of three observation aircraft to five fighters and one bomber. Just how influential the decision of the board actually was is difficult to determine: there were other factors which shaped the ultimate composition of the air arm. Among these, first in importance were the limitations on production in the United States and the nature of the demands made from overseas by the expeditionary force. Nevertheless, there can be no question of the proportionally immense weight of the board's initial decision, for subsequent alterations took on the character of modifications in the original program for production rather than fundamental changes in doctrine.

The influence exerted on aerial policy by the forces of the United States overseas did not wait upon the arrival of troops in large numbers. On 26 May 1917, while still in Washington, Gen. J. J. Pershing assumed his duties as commander-in-chief of the American Expeditionary Force. That same day he appointed Maj. T. F. Dodd of the Signal Corps to his staff as aviation officer of the Air Service, AEF. Major Dodd's appointment marked the establishment of an Air Service as an organization quite independent of the Signal Corps in the overseas theater, although the relationship of the Signal Corps and its Aviation Section remained unchanged in the domestic establishment.[14] Shortly after General Pershing and his newly formed staff arrived in Europe to lay the groundwork for the expeditionary force, Maj. William Mitchell, already in France as an observer, presented the Chief of Staff, AEF, with a study proposing the organization of an air force. Mitchell's plan divided aviation into two broad categories. The first consisted of

---

13. *Congressional Record, 55,* Pt. 5, 65 Cong. 1 Sess., 14 July 1917, 5131. The whole debate, pp. 5104–39, gives a résumé of aviation in the United States.

14. "The Origin of the Air Service, AEF," and research data compiled prior to preparation of study for Vol. *1,* chap. iii, Hist. A.S. AEF, 1919, NA, WWI Orgn. Records, A.S. Hist. Records Box 300.

squadrons serving the divisions, corps, and armies of the ground force and was to be attached just as units of other arms and services, such as artillery or signal troops, were attached. The second group comprised an entirely different type of force. Major Mitchell substantiated his recommendations regarding this group with a reference to French requirements.

> Based on the theory that no decision can be reached on the ground before a decision has been gained in the air, the French General Staff has requested that in addition to the aviation units which form a part of the American troops coming to France, there be organized a number of large aeronautical groups for strategical operations against enemy aircraft and enemy materiel, at a distance from the actual line. These units would be bombardment and pursuit formations and would have an independent mission very much as independent cavalry used to have, as distinguished from divisional cavalry. They would be used to carry the war well into the enemy's country.[15]

Mitchell's conception of the independent strategic mission of air power as analogous to strategic use of cavalry raiders cannot fail to raise speculation as to what might have been the evolution of aerial doctrine in the United States had the Cavalry rather than the Signal Corps served as the foster parent of aviation. Whatever "might have been," Major Mitchell's study thrust the problem of the proper organization of aviation to the forefront. Six days after receiving Mitchell's plan, General Pershing appointed a board of officers to determine the form and composition of the Air Service, AEF. The membership of the board is revealing: it included four Signal Corps representatives, one Cavalry officer and one Field Artillery officer, each bringing to the board's deliberations a somewhat different idea of the role of air power.

The prevailing French military doctrines regarding the role of the air weapon as well as the views of Major Mitchell helped the six officers in framing a final report. The board regarded as "a cardinal principle in warfare" the assumption that "a decision in the air must be sought and obtained before a decision on the ground can be reached." To this end the board recommended that the

15. *Ibid.* Memo, Maj. W. Mitchell to Chief of Staff, AEF, 13 June 1917, quoted in "Air Service Programs, 1917," study for chap. v of Hist. A.S. AEF, NA, WWI Orgn. Records, A.S. Hist. Records Box 300. For a statement of doctrine issued over Mitchell's name, see *General Principles Underlying the Use of the Air Service in the Zone of the Advance, AEF* (Hq., AEF, 30 Oct. 1917).

composition of the Air Service follow the program already suggested by the French. The French plan stipulated a strategic force of 30 bomber groups and 30 fighter groups for one element of the air arm and for the other a service force of a size determined entirely on a troop basis to take care of the ground arms. The board prepared copies of its recommendations for General Pershing's signature in the form of cables to the War Department. Apparently the cables never left France.[16] The composition of the Air Service, AEF, remained unsettled until 11 July 1917. On this date Pershing formally approved the General Organization Project, a comprehensive plan for the AEF as a whole which had been drafted by the Operations Section of his staff. The program thus authorized required an expeditionary force of a million men and provided for aircraft in conjunction with ground troops only. In all there would be 59 squadrons of tactical aircraft with the field armies. Strategic aviation found no place in this initial program of the AEF.

The strategic role of aviation, nevertheless, was not to be ignored. In August Mitchell, by then a lieutenant colonel, once again proposed an organization for the Air Service similar to that requested by the French. The 59 squadrons contemplated in the approved program of the AEF consisted of 39 for observation, five for bombing, and 15 for pursuit. Mitchell proposed in addition to the tactical force a strategic force of 201 squadrons divided into 41 for observation, 55 for bombardment, and 105 for pursuit.[17] He must have been persuasive: during October 1917 his proposal became the official program for aviation in the AEF. Program number one, as it was subsequently called, represented a total of 260 squadrons: 120 pursuit, 80 observation, and 60 bomber. This force included both strategic and tactical aircraft.

During the early months of 1918 the French General Staff reiterated its earlier contention, declaring that an increase in units for pursuit and bombardment would be "the most important reenforcements" required by the Allies. In April 1918 a subsequent program for the AEF reflected something of this influence; its authors asked for a total of 120 squadrons: 14 pursuit, 50 observation, and 56 bomber. The next authorized program, in June

16. See above, p. 46 n. 14. AEF aviation cables are on file in the National Archives, but they are difficult to use as they are still in corded bundles. A search conducted by an Air Service officer working on a study (mentioned above, p. 46 n. 14), immediately after World War I revealed no trace of these cables, nor was a contemporary search more successful.

17. See above, p. 43 n. 7.

1918, was even more clearly aligned with French thinking: 120 squadrons for pursuit, 40 for observation, and 101 for bombardment.[18] From these figures it might appear that General Pershing's staff had accepted wholeheartedly the strategic air doctrines of the French.

Unfortunately, a wide disparity existed between the official programs of the AEF and the number of aircraft actually on order in the United States. In April 1918 the aircraft on contract included 2,000 pursuit, 1,050 bomber, and 8,000 observation.[19] Not only was the number of aircraft on contract out of balance with the schedules of the AEF, but the production actually achieved in the United States also exerted a compelling influence on the planning of programs in the AEF. The final composition of the Air Service anticipated in July 1918 comprised a total of 202 squadrons: 60 pursuit, 101 observation, and 41 bomber units. While this distribution was probably based on a realistic attitude concerning the availability of aircraft rather than upon any overt change in the doctrines of air power, the Chief of the Air Service, AEF, approved the program, and thus it could scarcely fail to be influential in determining priorities on production in the United States. The following table summarizes the successive programs of the Air Service, AEF, and shows the changes in the composition of the proposed force:

### NUMBER OF SQUADRONS

| Date of Program | Pursuit | Observation | Bomber | Total |
|---|---|---|---|---|
| 17 Oct. 1917 | 120 | 80 | 60 | 260 |
| 9 April 1918 | 14 | 50 | 56 | 120 |
| 6 June 1918 | 120 | 40 | 101 | 261 |
| 29 July 1918 | 60 | 101 | 41 | 202 |

After the close of the war officers in the Air Service were sometimes inclined to point to the final program with its ratio of approximately two bombing squadrons to three fighter squadrons and five observation squadrons as evidence of a failure in the high command to appreciate strategic aviation. One commentator even declared that the final program "clearly considered the observation

18. "Formulation and Distribution of Programs," study for BAP Hist., Table XII, Sept. 1919, NA, BAP Hist. Box 9, 334.8 Overseas Missions.

19. *Ibid.* Memo, Col. H. H. Arnold, April 1918. The seemingly abnormal number of observation aircraft is probably accounted for in part by the inclusion of day-bombers in this category. The same aircraft served for both tactical bombing (day-bombing) and observation. The limits of performance of this type precluded its use in the strategic role contemplated by Mitchell, who visualized a force in which heavy, long-range night-bombers predominated.

squadron as of first importance" and must have stemmed from a
belief that aviation was "not to be used as a fighting arm except
for observation defense." [20] Some officials in the upper echelons of
command in the AEF may have felt this way, but there is abundant
evidence to show that the program of July 1918 was based on the
possibilities of production rather than upon any well-thought-out
policy.

Whether the program reflected the influence of production on
planning or the influence of planning on production, in the final
analysis experience in combat with aircraft of the several functional types would provide the basis for shaping most postwar
thinking about air power. Just how successfully the United States
met the problem of getting aircraft of these various functional
types to the fighting front remains to be told.

The pages above point out that exploitation of the air weapon
depended upon two critical factors: doctrine and equipment. Doctrine alone would win no wars. Even before they had been able to
refine the army's views on the mission of the air weapon or determine the proportions of the several functional types required,
officers of the Signal Corps had to decide upon specific models of
aircraft to fulfill the mission settled upon. When evaluating in
retrospect the selection of types which was eventually made, one
must not forget that the officers concerned had to choose their tools
before learning the nature of the job to be done.

Certainly no one in Washington could pretend to be fully informed of the task ahead. Ten days after the United States entered the European conflict the Chief Signal Officer called a meeting of all the Allied military attachés present in Washington and
told them that the Signal Corps would welcome foreign aviation
missions to this country. Hitherto Allied representation had been
more diplomatic than technical. But now the Signal chief believed
that special aviation missions staffed with men trained in design
and construction as well as pilots with tactical experience would
be essential in setting production on its way in this country.[21] The
Signal Corps's request for technical missions directly through the
good offices of the military attachés is significant. Earlier interna-
tional dealings, in the months of war from August 1914 to April
1917, were conducted almost exclusively through normal diplo-

20. *Ibid.*, cf. Benedict Crowell, *America's Munitions; 1917–1918* (Washington,
GPO, 1919), p. 254.

21. R. M. McFarland, "Foreign Missions to the United States," 21 July 1919, NA,
BAP Hist. Box 10, 336.91 Foreign Missions.

matic channels. This policy was probably, in part at least, a result of President Wilson's scrupulous attempts to keep the nation free from the slightest odor or tint of unneutral conduct.[22] Whatever may have been the diplomatic value of the policy, it contributed to the isolation of the United States from the rapidly evolving European doctrines of air power. The Signal Corps, however, appealed for pilots, designers, and engineers directly to the attachés "without proceeding through the regular diplomatic channels." Soon afterward a large number of French, Italian, and British officers and engineers came crowding to Washington and reported for duty directly to the chief of the Signal Corps. To handle this assortment of liaison officers a special organization was created directly under the chief.[23] The newly established office is worth noting, first, because it was relatively high in the chain of command and with the ear of the responsible official and second, because it represented an abnormal channel of information between Europe and the United States. The regular channel was of course the system of military attachés.

While the Allies were assembling their aviation missions in Washington (a slow process inasmuch as transatlantic passage in those days meant a steamship journey), another less official liaison group appeared. Agents for foreign manufacturers with their eyes on the potentially great needs of the Signal Corps and the relative insignificance of the aircraft industry in the United States turned up at every door with propositions. Most of these agents were willing to reveal trade secrets and ready to grant manufacturing concerns in this country liberal license rights on advanced types of aircraft in use at the front in Europe—in return for the payment of large royalties. Many advantages, to be sure, were likely to stem from direct contact with the aircraft manufacturers in Europe. They were better informed as to the tactical requirements of the war, and they were already employing techniques of mass production rather than job-shop methods of manufacture. But these advantages were offset to some extent because the European manufacturers were, for the most part, represented in the United States

22. For another instance of Wilson's adverse influence on technological preparedness in his anxiety to remain neutral, see AAF Historical Study No. 50, "Materiel Research and Development in the Army Air Arm; 1914–1945," p. 13, 1946, AAF Archive.

23. Report of Chief Signal Officer, *Annual Reports of the War Department*, 1918, *1*, 1074; and *The Signal Corps and Air Service . . . 1917–1918*, War Department Document No. 1109, 1922, Army War College Historical Section Monograph No. 16, pp. 40–1.

by brokers or commission agents and not by engineers. Moreover, financial embarrassments would almost certainly arise if the United States became involved in a series of royalty claims with individual foreign concerns. Accordingly an early decision was made requiring all inter-Allied exchanges of technical information to flow through official channels.[24] Whether the financial gain secured by this policy was sufficiently great to make up for the loss of direct technical relationship between manufacturers on both sides of the Atlantic is impossible to determine. Whatever the merits of the decision, it served to emphasize the importance of liaison on engineering matters between foreign governments and the United States.

Early in May 1917 an official of the NACA, concerned over the absence of planning for production in the United States, urged the Chief Signal Officer to order a group of engineers to Europe to secure the latest information about tactical aircraft. The Chief of the Signal Corps was naturally reluctant to comply since he had already invited the Allies to send missions to this country for that particular purpose, so the proposal was discarded. The foreign missions to the United States, however, proved less helpful than had been anticipated. The various officers and engineers sent to Washington were out of touch with the most recent developments along the front. Their information, based chiefly on personal experience, was weeks or even months out of date. When this became evident, the Chief Signal Officer at last approved a plan to send a technical mission to Europe.

Although the mission was approved on 15 May 1917, its members did not actually sail until the middle of June. Considering that every delay at that time played havoc with the nation's entire program for producing aircraft, the time between the date the mission was approved and its actual departure appears unreasonably long. Had the mission sailed a month sooner, the composition of the Air Service, AEF, would have been completely different. By arriving in France before General Pershing and his staff, the members of the aviation mission could have made their technical decisions solely in the light of French and British experience. Their decisions would then have reflected the aerial doctrine of those armies rather than the policies of Pershing's staff. The speculation is not an entirely idle one: the disparity between what was actually done and what might have been done points up the power

24. "The Aircraft Production Board," *Proceedings of the Acad. of Pol. Sci., 7,* No. 4 (Feb. 1918), p. 109. See also Testimony of Col. E. S. Gorrell, *War Expenditures,* House Hearings, 66 Cong. 1 Sess., Serial 2, Pt. 6, 4 Aug. 1919, p. 211.

of technical decisions to affect doctrine and the influence of doctrine on weapons.

Colonel R. C. Bolling, an officer who in private life had at one time served as general counsel for the United States Steel Corporation, was the leader of the mission leaving for Europe. An early aeronautical enthusiast, he recognized the importance of aircraft in warfare and had taken a leading role in organizing the first National Guard Aero Company in 1915. Besides Bolling, two aeronautical engineers from the Signal Corps, two officers representing aviation in the navy, and two civilians (both from the automotive industry) representing the point of view of production made up the total membership of the Bolling Mission, as that group came to be known. Nearly a hundred specialists, technicians, and mechanics accompanied the mission to acquire training in shop practices used in manufacturing aircraft and engines.[25]

The Bolling Mission set off to Europe armed with broad credentials authorizing it to consider almost every problem touching air materiel. The main objective of the mission was, however, to secure information regarding the latest designs of aircraft, which could be used as the basis for production in the United States. Just how vague the mission's directives were appears in the following cable sent by the Chief Signal Officer to Colonel Bolling a whole month after the group sailed from the United States: "Inform us as to types and numbers airplanes and engines we should concentrate on. . . . Want to know approximately type airplanes which will be wanted eight and twelve months from now." [26] Messages such as this certainly added nothing to the scope of the mission. Neither did they clarify its objectives. They merely uncovered the confusion and lack of appreciation of the nature of the problem prevailing in military circles in Washington. Moreover, the exact line of demarcation between the authority of the Bolling

25. Memo, Capt. J. L. Ingoldsby, Chief, Hist. Sect. A.S., 3 June 1919, NA, BAP Misc. Hist. Box 1, 321.9 A.S., Training; and Sweetser, *The American Air Service,* pp. 64–5. Gorrell, writing 23 years afterward, puts the number in the mission at 12; he probably considered some of the technicians as members. See E. S. Gorrell, *The Measure of America's World War Aeronautical Effort* (Northfield, Vt., Norwich University, 1940), p. 3. At the time of the mission's departure for Europe, Bolling was still a major. The two other army members, E. S. Gorrell and V. E. Clark, were both captains. Subsequently Bolling and Gorrell were promoted to full colonels, and Clark became a lieutenant colonel. Bolling, as the head of the mission, carried credentials both as an officer and as a civilian, so the question of rank was probably of little significance in his negotiations. Where exact rank at a given time is unknown, highest rank is cited.

26. Transcript, cable, Squier to Bolling, 17 July 1917, NA, BAP Hist. Box 20, 452.1 Airplanes, General.

group and the authority of officers in the Air Service, AEF, remained in doubt. Bolling's objectives were supposed to be "technical," but it soon became apparent that neither technical nor engineering decisions could be made without reference to tactics or
strategy.[27]

Two considerations enormously increased the difficulties of the
Bolling Mission. First, the mission had sailed from the United
States before the General Staff approved the program for aviation drawn up by the Joint Army-Navy Technical Board. Since it
was the board's program, based on the quantitative but not the
doctrinal content of Ribot's cable, which determined the composition and hence the doctrine of the Air Service, Bolling left the
United States without an authoritative statement of policy on the
nature of the role anticipated for the aviation units for which
the mission was to select aircraft. The members of the mission had
time to become familiar with the contents of Ribot's cable, but official approval took place after they sailed. In the second place,
when the mission arrived abroad, the Allies themselves after nearly
three years of war failed to agree upon a common doctrine to
present to the United States as the basis for a large-scale program
of production. Air officers among the Allies were in substantial
accord, but there appeared to be a diversity of opinion among the
high commands of Italy, France, and the United Kingdom.

When the French urged the formation of an independent or
strategic bombing force, the Royal Flying Corps resisted the move,
not because of any objection to the idea in principle but rather
because of an acute shortage of aircraft. At the time of the United
States' entry into the war, the British had 50 squadrons in France
—21 squadrons for service with the ground forces, 27 squadrons
of fighters, and only 2 squadrons of bombers. Later, after the
Germans had used giant Gotha bombers to demonstrate the utility
of strategic air power by long-range attacks on London, the Air
Board of the British government recommended a program to
organize 40 squadrons of long-range bombers. General Sir Douglas Haig, representing the point of view of the ground forces, opposed the proposition, fearing lest it would be accomplished only
at the expense of units cooperating with the ground arms. Thus,
although there were many in the United Kingdom who were thor-

27. "The Bolling Aeronautical Commission," study and research data compiled
for Vol. *I*, chap. iv, Hist. A.S. AEF, NA, WWI Orgn. Records, A.S. Hist. Records
Box 300.

oughly interested in the possibilities of strategic bombing, when the Bolling Mission arrived in London the British had been unable to settle upon a program for large-scale strategic bombardment.[28]

The views of the French have already been mentioned. In a situation similar to that in Britain, French production was unable to meet both the requirements of the armies and those of a strategic force. To overcome this difficulty the French contemplated using the first fruits of American productive capacity, the 4,500 aircraft specified in Ribot's message, to form a strategic force. Some of the circumstances which subverted this intention have been discussed earlier in this chapter. Even though Ribot's cable failed to carry its full implications on aerial doctrine, the French left no doubt in the minds of the members of the Bolling Mission as to the relative importance of the strategic air arm.

Italy was the only country in which the Bolling Mission found an effective long-range bombing program actually underway. On the Austrian front, Italian bombardment sorties had been so successful that one general officer was willing to sign a declaration of his confidence that "systematic, continuous, and scientific bombing" could effect an Austrian withdrawal in two weeks. The Italians had already concentrated as many as 250 Caproni bombers in a single raid. It was not mere numbers, however, which the Italians stressed but the importance of the doctrine that bombing should be "systematic, thorough, and consistent." [29]

By the end of July 1917, about a month and a half after sailing for Europe, the Bolling Mission had visited all the Allies and secured information upon which to base a report. At a conference attended by representatives of the French, British, and Italian governments, as well as the aviation officers of General Pershing's staff, the members of the mission prepared a cable to Washington describing in detail the specific types of aircraft they had chosen for production in the United States. Physical samples of each aircraft selected were dismantled and shipped across the Atlantic to help familiarize the prospective manufacturers with the foreign models. Although the United States participated in World War I for a total of 19 months, it was not until after the first five of these

28. H. A. St. G. Saunders, *Per Ardua* (London, Oxford University Press, 1945), pp. 219–20. See also Col. R. C. Bolling to H. E. Coffin, Chairman, Aircraft Board, 15 Oct. 1917, NA, BAP Hist. Box 21, 452.1 Caproni Contract; and H. A. Jones, *The War in the Air* (5 vols., *2–6;* Oxford, Clarendon Press, 1928–37), Vol. *6,* chaps. iii, iv.

29. Bolling to Coffin, 15 Oct. 1917, NA, BAP Hist. Box 21, 452.1 Caproni Contract.

had passed that the initial selection of designs was made and detailed planning for production could begin.[30]

The formal report of the Bolling Mission represented something of a middle course between the extreme views of the exponents of strategic bombardment on the one hand and the advocates of army-cooperation or tactical aviation on the other. Although the directives Bolling had received did not require or even expect the mission to formulate doctrine, it was probably inevitable that the report of the group should involve at least implicit consideration of principles. The Bolling report laid down a pattern for production: first, training aircraft; second, aircraft for use "strictly in connection with the operation of American forces in the field"; and a third category best described in the words of the report itself.

> After these first two considerations comes the American program of putting into the field next year air forces in excess of the tactical requirements of its army in France. It is greatly desired that the United States shall do this. Such air forces should consist of fighting airplanes and bombers.

The report, as drafted, established a time schedule for production. In the United States, third-place mention of the strategic force was apparently taken to mean that it was third in order of relative importance. As a consequence of this interpretation, officials did not treat bombers as a part of the immediate manufacturing program, and the composition of the Air Service with the AEF in France was modified accordingly.[31]

British experience, mentioned above, had already supplied ample evidence on the vital influence of the achievements or failures of production in shaping air doctrine. The members of the Bolling Mission were probably unaware that the order of priorities inferred from their report would have a profound influence upon the historical evolution of aerial doctrine. The opinions of the individual

30. Gorrell, "What, No Airplanes?" See also Gorrell testimony, cited above, p. 52 n. 24.

31. Photostat of signed copy, Col. R. C. Bolling to Chief Signal Officer, 15 Aug. 1917, NA, BAP Hist. Box 10, 334.8 Bolling Report. A cable regarding the aircraft selections mentioned in this report had been sent to the U.S. at the end of July 1917. Apparently there was some confusion in decoding or transmission, for on 7 Aug. 1917 a portion of the cable was repeated. See also C. E. Hughes to the Attorney General, 25 Oct. 1918, hereafter cited as Hughes Report, which contains significant portions of the Bolling message. The Hughes report was widely reprinted at the time it first appeared. The version cited in this study appears in *Automotive Industries; The Automobile, 39,* No. 18 (31 Oct. 1918), 745 ff.

members of the mission, expressed in correspondence apart from the formal report, show that they had a stronger appreciation of the strategic role of air power than their report would suggest. Specifically, Colonel Bolling wrote of his conviction that both the French and the British had overemphasized fighters at the expense of bombers. By contrast Italian experience gave substantial proof of the profitable results to be derived from a strategic bombing force.

In the brief time available the Bolling Mission had not tried to evaluate all the factors entering into the success of Italian bombardment: the absence of effective Austrian opposition was one element to be considered, the early success of Italian manufacturers in producing a large number of bombers another. Whatever the factors may have been, the scope of Italian bombing operations impressed the members of the mission. These operations along with the opinions of the British Air Board and the French General Staff combined to induce Colonel Bolling to favor bombing aircraft over the other functional types. Sometime after the report from the mission had gone to Washington, Colonel Bolling wrote to the man responsible for aircraft production in the United States: "I cannot strongly enough express my concurrence in your view that our great effort should be directed toward an enormous quantity production of bombing machines both for day and night." [32] This view of air power was certainly a great deal stronger than that expressed in the formal report of the Bolling Mission, giving credence to the thought that the members of the mission did not intend to give lowest priority to the independent bombing force. But the language of the report certainly seems to favor the opposite view when it says that the bombing units would be formed from "forces in excess of the tactical requirements" of the ground armies. Production had already been delayed five months while waiting for a decision regarding types. It must therefore have been apparent that the creation of an independent bombing force would suffer a critical delay if it also had to wait until the requirements for the ground forces could be met.

In view of the generally uncrystallized state of air doctrine in the United States except insofar as a policy had been read into Ribot's message, a large element of responsibility for the formulation of doctrine appears to fall upon the Bolling Mission. The members of the mission were responsible not only for their technical decisions but for their decisions on aerial doctrine as well, even if

32. Bolling to Coffin, 15 Oct. 1917, cited above, p. 55 n. 28.

the latter were implicit, incidental, or unintentional. Nonetheless, in making an assessment of the mission it should be noted that representatives of the Allied armies and of General Pershing's staff sat in with the Bolling group on the deliberations leading to the selection of types cabled to Washington. The mission's interim reports and the correspondence of individual members give evidence on their opinions when not influenced by the presence of staff officers of the Allies and of the AEF. The great detail with which Colonel Bolling reported opinions at variance with the implicit priorities established in the formal report of the mission gives some weight to the suspicion that the members of the Bolling group wanted to emphasize the doctrine of the independent bombing force regardless of the official declaration in their report.

As early as 29 June 1917, shortly after the mission first arrived in London, Colonel Bolling reported at length on the views of Gen. Sir David Henderson of the British Air Board. In a comprehensive discussion of the problem of aerial doctrine confronting the United States, General Henderson urged Colonel Bolling to abandon the attempt to regard aviation as a "balanced arm" established entirely in proportion to the number of troops in the ground forces. The general divided aviation into three categories: service or observation aviation, fighter aviation, and bombardment aviation. He believed that aircraft for observation should be procured on a troop basis—so many aircraft to serve so many troops. Fighters, the general felt, should be procured in quantities great enough to drive the enemy out of the air, preferably a three-to-one numerical superiority. General Henderson considered bombers in an entirely different light:

> Over and above the army machines and the fighters, there must be on hand a maximum number of airplanes that a country is able to produce to use against the enemy in bombarding him out of his position and cutting off his communications and destroying his sources of supply.

In the light of the subsequent lack of interest in the strategic air arm, this statement of doctrine is the more noteworthy for its clarity as well as for the early date at which it was expressed. The final phrase extended the vista of air power well beyond the range generally prevailing in military circles. The opinion is noteworthy for another reason as well: the man who expressed this view was a general officer, a representative of the British army. He concluded his plea to Colonel Bolling emphatically, saying, "The greater the number of bombarding machines, properly protected from attack of enemy machines, the shorter will be the duration of the war." [33]

In the absence of an arbitrary directive from Washington, the Bolling Mission was more or less free to shape air doctrine by the nature of its technical decisions and the emphasis of the implications for policy included in its reports to the United States. Nevertheless, despite the apparent attempts of individual members within the mission to influence the concept of the role of the aerial weapon as formulated in the United States, there appears to have been little change in the composition of the Air Service formally approved by the General Staff in June 1917.

The weight of the formal Bolling report with its concurrences by the staff of the AEF, as contrasted to the interim reports and letters of individual members of the mission, may have led to a decision in Washington to accept the composition of the force formulated by the Joint Army-Navy Technical Board as essentially sound. On the other hand, it may equally well be that once an aerial doctrine was formulated, albeit by implication, concerning the composition of the force planned for the Air Service of the AEF, officials in the Signal Corps were unwilling to make any changes despite additional information from abroad. Once production was begun, drastic alterations in policy would mean further serious delays. Moreover, in presenting a new program to the General Staff for approval, there was some danger that the opposition expressed earlier by the staff might on a second occasion prove more formidable. This was especially so since the hectic furor which had prevailed at headquarters in Washington during the first few weeks of the war had somewhat subsided.

The Bolling Mission had a remarkable opportunity to shape the content of and give direction to the doctrine of air power in the United States. To say that the mission failed to make the most of this opportunity is to cast no discredit upon the individuals in the group. Their assigned objective was to determine the best possible types of aircraft for production in the United States. The mission accomplished this objective. At the meeting of 31 July 1917 the Bolling group selected four major types of Allied aircraft for use by the AEF. They designated the British DeHavi-

33. Cited in "The Bolling Aeronautical Commission," study and research data compiled for Vol. I, chap. iv, Hist. A.S., AEF, 1919, NA, WWI Orgn. Records, A.S. Hist. Records Box 300. Gorrell's retrospective contention in 1941 that the British in 1917 had recommended against bombers as "of but little value" is not supported by the evidence of 1917, including his own reports. See above, p. 52 n. 24, p. 56 n. 30.

land, the DH-4, as the best aircraft for observation and day-bombing available for quantity production; as fighters they chose the British Bristol and the French SPAD, and the Italian Caproni was singled out to serve as a long-range night-bomber.[34]

Since there is no question of the integrity of the individuals involved, it may reasonably be assumed that they based their selections of aircraft on the best available technical and tactical data. How difficult it sometimes was to obtain such objective evidence will be shown in a subsequent chapter. But in this instance the decision to rely upon the British for three of the major types is attributable to several factors extraneous to effectiveness in combat or technical considerations. When the Bolling Mission arrived in London, the whole question of proprietary interests was at fever pitch. Allied manufacturers' agents had offered to license individual concerns in the United States at substantial fees. The War Department as the ultimate purchaser decided to deny all such payments in principle and to deal directly with the Allied governments. As soon as the Bolling negotiations began, the British government expedited the selection of types by saying in substance: we shall talk production now; let royalties wait until after the war. The French, on the other hand, insisted on working out the details of payment before initiating the production which would help win their war for survival.[35]

While the question of royalties probably contributed something to the mission's decision, there were some less tangible influences. According to the testimony of one military observer commenting on conditions in France immediately following the Bolling Mission's report, aircraft production was "terribly demoralized." Officers from the United States were reported as believing, possibly erroneously, that some individual manufacturers exercised so much political power in the Chamber of Deputies that they could determine the selection of aircraft for the French army. This lack of "grip at the top," which was expressed during the latter part of 1917 in idle aircraft factories or, even worse, in production of obsolete types, must be taken into consideration when weighing

34. Testimony of Col. E. S. Gorrell, *War Expenditures,* House Hearings, 66 Cong. 1 Sess., Serial 2, Pt. 6, 4 Aug. 1919, pp. 211 ff.; Gorrell, *The Measure of America's World War Aeronautical Effort;* and Sweetser, *The American Air Service,* p. 11. The complete reliance of the United States upon the Allied aircraft types selected by the Bolling Mission belies Lloyd George's charge that manufacturers in the U.S. refused to accept Allied designs since this would have been a reflection on American "inventiveness." See David Lloyd George, *War Memoirs* (6 vols.; Boston, Little, Brown, 1933–37), *5,* 451. See also Hughes Report, 25 Oct. 1918, p. 750–12.

35. See Gorrell Testimony, p. 211, cited above, n. 34.

the elements involved in the decisions of the Bolling group.[36]

The reports of the Bolling Mission did not elaborate or explain in detail all the elements contributing to the decisions on design. There was little time for explanations. Only 35 days elapsed between the day the mission reached England and the date of the final report. During that period, members of the mission visited factories, conferred with military leaders, and studied sample aircraft in three different nations. When the decisions of the meeting in July finally were made, the mission had completed its assignment. Its members did not return directly to the United States, but applied themselves individually to some of the many other tasks which were everywhere crying to be done in preparation for the arrival of the great hordes of soldiers from across the Atlantic.[37]

Officials in the United States generally assumed that the report received from the Bolling Mission would unleash the productive capacity of the nation to create aerial flotillas capable of sweeping the Germans from the skies. But during the spring and summer of 1917 the military authorities directed no effort toward establishing an effective organization to continue making the decisions begun by the Bolling Mission. Here is evidence of their failure to appreciate at that time the dynamic nature of aircraft design. Freezing design at one point for any prolonged length of time would result in producing obsolete aircraft, a liability in combat. When the Bolling decisions were made this was not fully understood if, as is doubtful, it was recognized at all. The realities of combat were to drive the point home all too clearly.

At the end of July the Bolling Mission with the unanimous vote of all the Allies selected the French SPAD as one of the two best available fighters in Europe. During the second battle of Verdun in August 1917, enemy innovations made the SPAD obsolete. Because of delays on the part of the French authorities, the sample SPAD which the Bolling Mission had ordered as the starting point for the production of fighters in the United States had not left France at the time it became obsolete.[38] And this instance was by no means an isolated one. Less than 48 hours after the Dutch air-

36. Transcript of testimony, Col. S. D. Waldon before special meeting of the Aircraft Board, 8 Feb. 1918, NA, BAP Executive Office Files, Box 65, 334.8 Aircraft Board. The BAP Executive Office Files are hereafter cited as BAP Exec.

37. The Bolling Mission formally dissolved 31 July 1917, and the members worked independently thereafter. The language of one report made on 4 Sept. 1917 does indicate, however, that it was compiled by more than one member. See NA, BAP Hist. Box 21, 452.1 Caproni History.

38. Gorrell, "What, No Airplanes?" See also Copy, Col. S. D. Waldon to Maj. H. S. Martin, 5 Sept. 1917, NA, BAP Hist. Box 25, 452.1 SPAD file.

craft designer Anthony Fokker first saw the principle of propeller deflector plates on a captured French aircraft, he worked out his subsequently famous synchronous interrupter-gear for firing machine guns through a propeller arc. While it is obvious that this device required somewhat more time to put into production than to conceive initially, the implication of a rapid flux in design is self-evident.[39]

By the end of the summer of 1917 Colonel Bolling had come to recognize the importance of changes in design as a factor in maintaining superiority in the air. He felt that the French and British had a decided advantage over the United States in their proximity to the fighting front. Changes evolved in actual combat could be hurried home and introduced as modifications on aircraft under construction, all within a matter of hours or days. To do the same for the United States involved delays of weeks and even months. The SPAD may have been in Colonel Bolling's mind when he subsequently wrote, "The whole complexion of the air situation may be said literally, and without exaggeration, to be capable of change overnight, due to some improved design . . ."[40] Thus the colonel recognized the importance of changes in design but failed to offer any specific recommendations concerning ways and means of meeting the situation.

Lieutenant Colonel Virginius E. Clark, one of the two aeronautical engineers in the Bolling Mission, also saw the constant and rapid changes in the design of aircraft on the front in Europe. Yet, like Colonel Bolling, he gave little indication that he appreciated the need for a system which would keep aircraft on the production line in pace with changes in design at the front. Before returning to the United States to assume a responsible position in the army's organization for aircraft engineering, the colonel reported the conclusions he had reached after studying the problem of the aerial weapon in Europe: "I believe that probably the most important influence on the conduct of war is the group of men held responsible for deciding which types will be built and supplied to the forces in the field."[41]

39. A. H. G. Fokker and B. Gould, *Flying Dutchman* (New York, Henry Holt, 1931), chap. xiii *passim,* esp. p. 124; and C. G. Grey, *The History of Combat Airplanes* (Northfield, Vt., Norwich University, 1941), p. 6.

40. Bolling message of 4 Sept. 1917, quoted in A. B. Gregg, "History of the Caproni Biplane," 1919, BAP Hist. Box 21, 452.1 Caproni History.

41. Lt. Col. V. E. Clark, "Final Report on the Present Status of Military Airplanes along the Western Battlefront of Europe," 12 Sept. 1917, NA, BAP Hist. Box 20, 452.1 Airplanes along the Western Front.

Colonel Clark saw the problem as a selection of types to be followed by production. His recommendations as to the method of making this selection confirm the suspicion that in his conception of the problem he visualized the choice as a single act rather than the first of a long series of decisions, each concerning a change in design necessary to keep pace with the enemy. The group of men responsible for making decisions as to types of aircraft should include, Colonel Clark believed, individuals representing a variety of points of view in such fields as operations in combat, maintenance at the front, design of aircraft, engineering and manufacturing. This attitude showed that Colonel Clark expected decisions to be made using information based on a wide range of activities, but the information was to come from the past experience of the men making the decisions rather than from any continuing system for the orderly accumulation of data.

Subsequent events were to demonstrate the futility of relying upon "experience" rather than continuously accumulating data from every quarter as a basis for decisions concerning changes in the design of aircraft. At the end of 1917 there were few in authority who appreciated this principle, and the Bolling Mission broke up, leaving behind it no more than a rudimentary organization to perform the function of gathering information (about enemy developments, new tactical applications, problems of maintenance, possible exploitation of advances in science and capabilities of production) and the function of making decisions (initial selection of new types, introduction of changes in design, modifications, and similar matters). During the remaining months of the war there gradually emerged an increasingly clear conception of the situation faced. As the problem found definition, organizations were improvised to meet the need.

The coming of war found the United States lacking in two elements essential to the successful exploitation of the aerial weapon: a doctrine or statement of the mission expected of the aerial weapon and a knowledge of the specific types of aircraft required to implement that mission. For good or ill, a doctrine was formulated from the misinterpretation of Ribot's cable and the obiter dictum of the Bolling reports. By the end of 1917 the necessary decisions as to types had been made, and manufacturers in the United States were in various stages of preliminary planning for large-scale production.

At the end of 1917 the problem of mass-producing aircraft might seem to have been half mastered. Actually the obstacles to production were only beginning to appear. Design of aircraft was not only unstable but far less static than the design of any other weapon military men had ever before encountered. Only slowly did the officials responsible for manufacturing aircraft in the United States come to recognize that the dynamic nature of design in the field of aeronautics imposed unusual difficulties on production. To overcome these obstacles would require an organization to assemble information and an organization to make decisions. Until these organizations could be contrived, General Pershing's armies marched without adequate air cover, and no American bombers threatened German industries.

*Chapter IV. Organization for Decision*

ON THE FIGHTING front the opposing air forces struggling for supremacy introduced innovations almost daily. Some, like the Fokker synchronizing gear, resulted in revolutionary changes. Others merely increased the margin of performance, giving one side or another the advantage in speed, rate of climb, ceiling, range, pay-load, or some other such performance characteristic. The course of technological advance was never a straight line; the process was a continual series of trials and errors. When one side appeared on the front with an aircraft of superior characteristics, the other had to devise design changes in both existing production models and experimental aircraft for effective countermeasures. Each change in design involved a decision. Each decision, based on the best information available, was a gamble that the design or alteration selected would prove to be one which, in happy conjunction with other such selections, would yield an aircraft superior to that of the enemy.

The agency charged with making decisions regarding design and production of air weapons during World War I was certain to carry a heavy burden of responsibility for the success or failure of aircraft at the front. And no matter what form this agency might assume, to ensure the production of superior aircraft would require an administrative mechanism skillfully devised to perform a complex series of differentiated functions. During 1917 and 1918 such a decision-making organization did evolve. But, in order to understand the evolution of this organization, a general knowledge of the over-all administrative structure for aviation in the United States is indispensable; and before any further attempt is made to analyze the decision-making problem it is necessary to digress somewhat to review the wartime relationship between the Signal Corps and several independent governmental agencies directing the production of aircraft.

When the United States entered the war in Europe, two factors complicated the problem of developing a force for aerial combat.

First, the existing organization within the Signal Corps was far too small and inexperienced to cope with the many difficulties suddenly confronting it; and second, independent agencies outside the corps were already performing some of the functions involved in expansion. Thus from the outbreak of war the Signal Corps faced not only a staggering problem of expansion but also the necessity of accepting responsibility for the activities of agencies over which it had little or no control. An appreciation of this situation goes far toward explaining, if not excusing, the failures which blighted the aviation program throughout the war.

The Aviation Section of the Signal Corps in the early months of 1917 consisted of a few dozen pilots, mostly in the field, and a mere handful of staff officers stationed in Washington. Upon this tiny group in the capital fell the entire burden of planning the aviation program of the nation. Since this small staff was obviously fairly overwhelmed with the task of planning for airfields, personnel, and training as well as with the myriad other details of the enterprise, it is not surprising that leadership in the matter of production fell into other hands—at first, those of the National Advisory Committee for Aeronautics (NACA).

Congress created the NACA in March 1915 to promote the growth of aviation in the United States. Its present-day function makes it the nation's leading organization for fundamental research in aerodynamics, but at the time of its creation, in the form of a rider on a naval appropriations bill, the NACA was but little more than its title implies, an advisory organization. Its powers and functions were sufficiently broad and vague to allow participation in almost any aspect of the development of aviation.[1]

When war approached, officials of the NACA showed commendable initiative in obtaining statistics on production from aircraft and engine manufacturers to use as a basis for planning. In the absence of any large-scale, comprehensive survey by the Signal Corps the NACA statistics on production were invaluable as a point of departure for all subsequent procurement. Laudable as this display of initiative may have been, it undoubtedly contributed to the cleavage which later developed between production and operations. The NACA was legally responsible to no one but the President and quite beyond effective control by the Signal Corps.[2]

1. For a revealing analysis of the NACA in relation to the development of military aircraft, see AAF Historical Study No. 50, "Materiel Research and Development in the Army Air Arm; 1914–1945," Nov. 1946, AAF Archive.

2. *First Annual Report of the National Advisory Committee for Aeronautics* (Washington, GPO, 1916), pp. 9–20.

Soon after the declaration of war the rather informal activities of the NACA in planning for production were supplemented by a somewhat more formal but nonetheless "advisory" group. In May 1917 the National Defense Council, itself an advisory organization, established an Aircraft Production Board charged with coordinating the designs for army and navy aircraft and engines in addition to cooperating with the military services to remedy difficulties in production. The Aircraft Production Board, led by H. E. Coffin, an executive from the automotive industry, picked up where the NACA group left off in an effort to rationalize the problem of aircraft production. The board, like the NACA group before it, was staffed with civilians whose only authoritative channel of command to the Signal Corps lay through the Secretary of War.[3]

The Aircraft Production Board recognized that efficient mass-production of aircraft would require standardization wherever possible. Standardization implied prior agreement between the army and the navy as to the types of aircraft to be produced. Therefore, the Joint Army-Navy Technical Board, described in an earlier chapter, was formed in May 1917 to achieve this agreement. The technical board, supposedly representing some of the best qualified aviators in the two services, was originally charged with selecting types of aircraft and making sure that designs were coordinated and common specifications drawn up. This function gave the technical board final control over the effectiveness of aircraft sent to the front, but the board was directly responsible to the secretaries of War and Navy rather than to the Signal Corps.

The Defense Act of 1916 endowed the Chief Signal Officer with responsibility for the development of aviation for the army, but in wartime the chief found that he was dependent upon organizations outside the scope of his authority.[4] This situation developed from the failure of the Signal Corps to take the initiative in organizing to meet the exigencies of the national crisis.

3. "The Aircraft Production Board," *Proceedings of the Acad. of Pol. Sci.*, 7, No. 4 (Feb. 1918), 104. The Aircraft Production Board was established by a resolution of the National Defense Council, 16 May 1917. See also A. Sweetser, *The American Air Service* (New York, D. Appleton, 1919), chap. iii, for a brief description of the board. There is no existing definitive study of the board, but see typescript, "History of the Bureau of Aircraft Production," 1919, Wright Field Hist. Office File. See also W. F. Willoughby, *Government Organization in War Time and After* (New York, D. Appleton, 1919), pp. 331–5.

4. National Defense Act, 39 Stat. 166, Sect. 13, 3 June 1916. The NACA *Annual Report, 1917*, pp. 16–17, accepts full responsibility for originating the plan to put production under an agency outside the War Department.

After five months of war, during which time responsibility for the program had remained uncertain and divided among several agencies, the Chief Signal Officer formed an Equipment Division as one of the operating organizations of the Aviation Section. This new division was to unify and coordinate the several organizations then rather loosely related in the program of aircraft procurement. To this end a member of the Aircraft Production Board was commissioned and transferred to head the new establishment. Far from unifying the organizations already in existence, however, creation of the Equipment Division only increased the diversity of authority: instead of passing into extinction the Aircraft Production Board continued to operate with diminished functions. Finally, in October 1917 Congress regularized the position of the Aircraft Production Board by redesignating it the Aircraft Board and giving it the power to advise the army and navy on matters of production. While this board enjoyed a greater legal authorization than its predecessor, the Congressional action had done little to unify control of the program.

Without an effective organization for making decisions, production of aircraft on a large scale failed to materialize. The promised squadrons which were to bring the Allies unquestioned superiority in the air were nowhere to be found. In fact, during the first three months of 1918, production could scarcely be said to have begun, let alone to have attained the goals established.

This shocking situation provoked a series of investigations, including one by a Senate committee and another by agents of the Department of Justice, which subjected the failures of the program to widespread popular criticism and led to important reorganizations.[5] A preliminary, half-measure reform in April which grouped the aviation activities of the Signal Corps under a single, semiautonomous head reporting to the Chief Signal Officer failed to achieve effective unification of command. During the following month President Wilson used the emergency powers granted him under the Overman Act and authorized a completely new establishment for aviation.

The President's Executive Order of 20 May 1918 created two separate organizations to carry out the wartime program. As an initial step, the President removed the Signal Corps from all fur-

5. For an analysis of the several aircraft investigations, see J. A. Beck, "Investigations," and data compiled in preparation of study, Sept. 1919, NA, BAP Hist. Box 8, 333.5 Investigations, General.

ther participation in the aviation program.[6] In spite of the many difficulties that had been encountered as a result of the lack of centralized control over the program President Wilson's executive order still did not unify production and operations under one head. Apparently the tradition or practice of separate control over production on the one hand and operations on the other proved too strong to break down. In place of a centralized agency within the War Department for all functions of aviation, the President created two coequal agencies. The Division of Military Aeronautics (DMA), under a military head, assumed responsibility for personnel, training, and requirements; the Bureau of Aircraft Production (BAP), under a civilian head, dealt with the problems of production.

The reorganization of May 1918 simplified the problem of controlling the program for aviation, but simplification did not mean solution. The opposing halves continued to find it difficult to resolve their opposite interests until late in August 1918 when President Wilson appointed the civilian head of BAP as a Second Assistant Secretary of War and designated him Director of the Air Service to supervise both BAP and DMA. Although this step had an important bearing upon the postwar organization of the air arm, it came too late to have any real unifying effect during the war. The man chosen to fill the position, J. D. Ryan of Anaconda Copper, spent the larger part of his incumbency traveling in Europe to familiarize himself with the situation there. The Armistice and the end of the war came before he could effect substantial reorganizations.

Whatever savings in administrative overhead and elimination of duplicate functions may have been accomplished by the reorganizations of May, the creation of two mutually independent bureaus did nothing to improve or simplify the problems involved in the selection of aircraft. The new bureaus had no sooner been established than conflicts occurred over decisions on designs. The directives of the new bureaus purposely left this point vague so that the proper division of technical decisions might evolve from experience.[7] After a year of war it would appear to have been

6. Report of Chief Signal Officer and Report of Bureau of Aircraft Production, *Annual Reports of the War Department*, 1918, *1*, 1075, 1407. See also AAF Historical Study No. 25, "Organization of Military Aeronautics; 1907–1935," p. 32, 1944, AAF Archive.

7. Report of Director of Military Aeronautics, *Annual Reports of the War Department*, 1918, *1*, 1383.

rather late to delay any further the determination of this most critical policy of the whole program.

The War Department recognized that the problem of responsibility for the selection of types of aircraft was of the utmost importance. In fact, the annual report of the Director of Military Aeronautics in 1918 singled out this question for special discussion:

> An early defect discovered in the reorganization developed when there appeared to be inadequate liaison between the Bureau of Aircraft Production and the Division of Military Aeronautics. One was responsible for the production of airplanes, the other for their operation and military efficiency. The method of selecting a type to put into production and the final decision whether any airplane produced was suitable for its military purpose or not, was undetermined. The situation of two sets of officials with equal authority in their respective fields of action, neither responsible to the other, at once demonstrated that neither could be held for the final production of an acceptable airplane for the front.[8]

Not until six months after the United States entered the war did an orderly method for authorizing changes in design appear mandatory to the military officials, and when they finally recognized this necessity, the need for a solution had become urgent. Remedial action of a permanent character was imperative if manufacturers in the United States were to avoid turning out obsolete aircraft which were worse than useless at the front. A series of wrong decisions about aircraft design might conceivably result in a loss of superiority in the air sufficient to lose the war. Nonetheless, the action taken was hasty, expedient, and haphazard at best. Existing organizations, each established for other purposes, were turned to tasks for which they were unsuited. Among the most important of these organizations was the Joint Army-Navy Technical Board.

The board, established originally to make final decisions as to types of aircraft and to coordinate the programs of the army and the navy, was supposedly endowed with authority because it reported directly to the two secretaries. But gradually it became apparent that "authority" was no substitute for full information. In August 1917 General Pershing had cabled from France that decisions of the Joint Army-Navy Technical Board regarding types

8. *Ibid., 1*, 1385; spelling slightly altered to conform with present-day usage.

## The Use and Misuse of the Joint Army-Navy
## Technical Board
## (*JANTB*)

Comparative diagram showing the *JANTB* in three stages of its existence: first as a design coordination board for the army and navy; later as a production coordination board between the BAP and the DMA; and finally as a general technical liaison agency for the army and navy air arms.

of aircraft "should not be considered against our recommendation as we believe that direct contact here is essential . . ." [9] Thus undermined, the position of the Joint Army-Navy Technical Board deteriorated steadily throughout the first half of 1918. The decline resulted in part from want of information and in part from the growth of other agencies better equipped to fulfill the decision-making function. The activities of the board during June 1918 amply illustrate just how far these two factors had gone in undermining the agency.

When the Chief of the Air Service, AEF, demanded superior aircraft in June 1918 to replace the unsatisfactory types being produced in the United States, the board proposed to send an engineering mission across the Atlantic for data on the designs sought by the AEF.[10] The Director of Military Aeronautics agreed to this proposal, since the Air Service "obviously" would be "at a great disadvantage" if required to fight with inferior aircraft. The Secretary of War rejected this double plea, because, as he noted, the Bureau of Aircraft Production appeared to have the problem of making decisions on designs for aircraft well in hand.[11] Officials in charge of production, it would seem, were determining the types of aircraft sent to the AEF. The board, supposedly empowered to make final decisions, found its authority gone.

The most revealing indication of the decline of the Joint Army-Navy Technical Board soon followed. Secretary of War Baker, who a little over a year earlier had created the board in conjunction with the Secretary of the Navy, now replied to a memorandum from the Chief of Staff, "I do not know what the Joint Army-Navy Technical . . . Board is . . ." [12] If the board had once held au-

9. Copy, Pershing cable No. 70, 1 August 1917, NA, BAP Hist. Box 20, 452.1 Bristol Aircraft. Pershing's specific mention of the JANTB decisions belies the stand of postwar authors who reduce the JANTB to an advisory organization.

10. Joint Army-Navy Technical Board to Secretary of War, 8 June 1918, AFCF, 334.7 Army-Navy Joint Boards. The alteration in function and decline in importance suffered by the JANTB are shown in two postwar publications which reflect the board's status at the war's end and neglect its earlier role: The General Staff, Plans Division Historical Branch, Monograph No. 3, *A Handbook of Economic Agencies of the War of 1917* (Washington, GPO, 1919), p. 19, gives the JANTB stature "in advisory capacity only"; Willoughby, *Government Organization in War Time and After*, p. 331, limits the JANTB even more seriously, erroneously reporting that it "never played any controlling part in the formulation or execution of an aircraft program." Sometimes JANTB is referred to as the Joint Army-Navy Technical *Aircraft* Board.

11. First ind. to correspondence in n. 10, above, Adjutant General's Office to JANTB, 3 July 1918; see also Memo, Director of Military Aeronautics to Chief of Staff, 20 June 1918, AFCF, 334.7 Army-Navy Joint Boards.

12. Secretary of War to Chief of Staff, 15 June 1918, AFCF, 334.7 Army-Navy

thority because of its close contact with the secretaries, that proximity now appeared to be somewhat diminished. In actual fact the board had not so much lost authority as shifted its function. Created to coordinate decisions on design for the army and navy, the board had been pressed into an entirely different role. The tug of war which made an authoritative reaching of decisions so necessary was not between the army and the navy but rather between the officials controlling production of aircraft as against the officials controlling the use of aircraft in combat. While the officials governing production (BAP) strove to turn out aircraft in quantity, those in charge of operations (DMA) were concerned with quality, that is, aircraft equal or superior to those of the enemy.

Membership on the Joint Army-Navy Technical Board represented the army and navy rather than production and operations. The ineffectiveness of this representation gradually dawned on the officials who were hampered for want of authoritative decisions on design. In one instance, the board had attempted to cancel production of the three obsolescent models of aircraft which happened to be precisely those upon which the BAP pinned most of its hopes for record-breaking, large-scale manufacture. From the point of view of the BAP, any decision to cut off production, even if intended to prevent the manufacture of obsolete aircraft, was apparently considered unthinkable. It was only by a "lucky chance," the Director of Production reported, that the board had been deterred from recommending cancellation of the three models, and by June 1918 he had come to believe that the BAP should have representation on the board to "stop them putting through recommendations having to do with production." The language of this comment reflected the attitude of a typical production-minded official toward agencies making qualitative alterations which impaired quantitative results. Adding BAP representatives to the board might not have stopped the changes in design to the extent expected by the BAP officials, but officers representing production as well as design certainly would man the board more effectively for the task facing it than did the original army-navy membership. Production-minded representatives finally joined the Joint Army-Navy Technical Board during July 1918.

The discussions about the board during June and July led the Director of Production to suggest that a control board be set up immediately under the directors of DMA and BAP to unite the two

---

Joint Boards. The secretary called it "the Joint Army-Navy Technical Aircraft Board."

halves.[13] Just how such a board would be any more effective than
the reconstituted joint board he did not explain, but the suggestion
is an indication of the trial-and-error approach government officials used in meeting problems of administrative organization
for producing aircraft. Equally revealing is the implication that
the differences pulling the DMA and BAP apart could be resolved
by officials representing the two groups. The emphasis seemed to
be on the status of the individuals suggested for membership or a
reliance upon personalities rather than upon the administrative
soundness of the system. Sometime before, the Assistant Director
of Military Aeronautics, Col. H. H. Arnold, had described the
members of the Joint Army-Navy Technical Board as "the only
practical, theoretical and analytical men in the Army and
Navy . . ." The members were, he said, men "especially qualified
to pass on all designs and types before they are put into production." [14]

The reliance on "qualified" individuals which had prompted the
formation of the Bolling Mission still prevailed in the middle of
1918. Nevertheless, the futility of decisions made by boards not
constantly supplied with a stream of information was fast becoming apparent. No matter how high ranking or well qualified members of a board might be, if they lacked a flow of up-to-date information from the front as well as from industry their decisions
would run the risk of being sterile. That is what happened to the
Joint Army-Navy Board. By the time the war entered its final
month the board found itself superseded by better informed and
better organized agencies within the DMA and BAP. The members voted to dissolve. The War Department rejected this move
in the expectation that the board would provide a useful administrative device for "joint consideration of aircraft matters in which
both . . . of the Services are interested." [15]

The agencies of BAP and DMA which assumed the joint board's
function of decision-making by reason of their greater effectiveness
had achieved this superiority over the JANTB only gropingly.
Soon after the reorganization, the authorities of the coequal DMA
and BAP, saw the dangers inherent in the uncertain nature of
their relationship which President Wilson's executive order left

13. M. W. Kellogg to W. C. Potter, 13 June 1918, and JANTB to Secretary of
War, 24 July 1918, AFCF, 334.7 Army-Navy Joint Boards.

14. Memo, Asst. Director of Military Aeronautics to Director of Military Aeronautics, 20 July 1918, AFCF, 334.7 Army-Navy Joint Boards.

15. W. C. Potter to C. W. Nash, 3 Sept. 1918, and reply, 11 Sept. 1918, AFCF,
334.7 Army-Navy Joint Boards.

*Diagram of World War I Aircraft Production Control Agencies*
*Showing Variations in Channels of Command and Liaison*
*in Five Successive Phases*

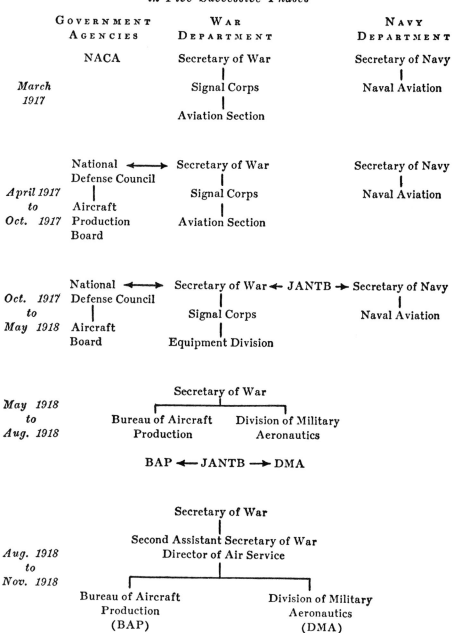

undefined. Responsibility for results without authority to control the means could not fail to breed trouble, and this was borne out in the conflict between the two organizations over the matter of design changes. On the one hand, officers of the BAP, in their anxiety to secure quantities of aircraft, had shown a tendency to favor stabilization of design, for it was primarily the high rate of change which prevented the United States from sending large numbers of aircraft to Europe during the early months of 1918. On the other hand, this unwillingness or inability of the authorities controlling production to recognize how important it was to keep the performance of the army's aircraft superior to that of the enemy by changes in design troubled the Director of Military Aeronautics more than any other single problem. A bare six days after the reorganization of May had gone into effect, Colonel Arnold, as Assistant Director of Military Aeronautics, expressed this fear when he complained that coordination of users (DMA) and suppliers (BAP) was inadequate. He touched the core of the problem in saying, "The Division of Military Aeronautics must control the determination of the design of the equipment with which it is to operate." [16]

To assign responsibility for design to the Division of Military Aeronautics in a formal directive was all very well, but to try to function in line with that directive was quite another matter. Each model in production represented literally thousands of technical features. Placing responsibility for design with DMA meant in practice requiring its officials to decide among a series of possible variations for each one of the critical features comprising a given aircraft. Since the problem was unavoidably technical, solutions were devised by the engineers who put aircraft into production rather than by officials who encountered the aircraft as a finished product. The experience of the industrial world was repeated in the military establishment: technical decisions tended to gravitate into the hands of the engineers actually working on the project regardless of where responsibility was placed by directive.

During the months when the Equipment Division of the Signal Corps was responsible for the development of aircraft, there had been two agencies within the division concerned with design. The Airplane Engineering Department held responsibility for develop-

16. Memo, Asst. Director of Military Aeronautics to Director of Military Aeronautics, 6 June 1918, quoted in AAF Historical Study No. 25, p. 35. See also *Aircraft Production in the United States*, Senate Report No. 555, 62 Cong. 2 Sess., 22 Aug. 1918, *passim*.

ing "new and advanced models," and the Production Engineering Department was responsible for expediting production, preparing drawings, and putting into large-scale manufacture the designs once they were perfected by the Airplane Engineering Department.[17] But the production engineers began to make changes in design. At first the changes were minor ones introduced to speed assembly, but they gradually became more and more important until the Production Engineering Department had, in the eyes of the chief of the Airplane Engineering Department, actually seized "control of airplane design itself." [18] The incident is interesting, because it shows how decisions on design tended to concentrate in the hands of production engineers even where directives clearly indicated the division of responsibility.

If the Equipment Division had had trouble in keeping decisions on design out of the hands of production engineers within its own organization, it is clear that the problem was even more difficult for the authorities of the DMA who were coequal with but nevertheless entirely separate from the BAP. As long as the DMA tried to reach decisions on the basis of the opinions which officers of its Technical Section held regarding the designs submitted by the BAP, the two agencies ran into many difficulties.[19] Colonel Arnold's insistence, in the first week after the reorganization took place, on the necessity of having the DMA control designs underlined his recognition of the nature of the problem and the need for a definition of responsibilities. Still, a practical solution remained to be found.

Under the reorganization of 20 May 1918 the problem of final authority for decisions respecting designs narrowed down more or less to the problem of defining the relationship of a few subordinate agencies within the two major and coequal branches, agencies such as the Technical Section (DMA) and the Engineering Section (BAP). Representatives of the two organizations struggled all

17. Signal Corps, Equip. Div., General Memo No. 88, 18 March 1918, presented as Exhibit A in "Organization and Operation of Bureau of Aircraft Production," 2 Aug. 1918, NA, BAP Misc. Hist. Box 1, 026.4 BAP.

18. Interview with Lt. Col. V. E. Clark, cited by J. A. Beck, "Investigations," pp. 232–5, Sept. 1919, NA, BAP Hist. Box 8, 333.5 Investigations, General. As head of engineering Clark was certainly not entirely disinterested, of course, when he critized production attempts to absorb design responsibility. See Capt. H. H. Blee, "History of . . . Airplane Engineering Division . . . ," 15 Aug. 1919, Wright Field Hist. Office Files.

19. Some indication of the Technical Section's conception of its responsibilities at about this period can be gathered from an organization chart of DMA prepared sometime before 27 July 1918, NA, OCAS Box 170, Charts, DMA.

during the first half of July to arrive at some acceptable working arrangement. By the middle of the month the two parties had reached what they believed to be a satisfactory solution. The Technical Section (DMA) and the Engineering Section (BAP) were located in one place physically, and the two organizations were ordered to report through a single head. In a few days this plan proved unworkable. Just as officers in the old Equipment Division of the Signal Corps had learned, leaders of the DMA and BAP discovered anew that merely placing the two conflicting agencies under one head would not cure the functional troubles lying behind the conflict.

Fortunately for the success of aircraft production this initial attempt at cooperation and agreement was not without its benefits. The same order which directed the Technical and Engineering sections to report through a common head introduced two other changes in the existing organization. First, it instructed the Director of Military Aeronautics to establish a Testing Section which would provide an objective basis for decisions on the designs developed by engineers of the BAP. Second, it established a central information agency to provide both the BAP and the DMA with information on which to base decisions.[20] Thus, although the attempted solution by means of joint operation failed, some progress had been made with the establishment of the Information and Testing sections to provide factual data upon which to base decisions.

By the end of July 1918 it was possible for officials of the DMA and the BAP to reach a much more effective agreement. They scrapped the idea of a single head for the Technical and Engineering sections and instead worked out a careful division of functions. Most significantly, they assigned to the Technical Section of DMA the specific responsibility of preparing specifications for performance, that is, objectives toward which engineers of the BAP could strive. The Technical Section, "as the direct representative of the user," was also empowered to establish the priorities for the development of new types and alteration of existing models. The DMA was to send technical representatives to tactical units to gather "first-hand information" at the front. Similarly, the new agreement authorized the transfer of some technically competent individuals to the recently formed Information Section to ensure a

20. Memo, W. C. Potter and Maj. Gen. W. L. Kenly to Chief of Staff, 16 July 1918, NA, BAP Exec. Box 81, 402.1 Technical Data.

continual flow of technical and tactical data from the operating units. Officials of the DMA and BAP conferring together were to make the decisions to begin manufacturing any given aircraft model, but the Technical Section retained the power of veto over changes in production.[21]

In short, the new agreement amounted to nothing more than an effort to provide the DMA with agencies and procedures to carry out a responsibility that had long since been assigned, on paper at least, to the Director of Military Aeronautics. Its real significance was twofold. In the first place, by assigning responsibility to the Technical Section for the preparation of performance specifications it was possible to escape from the futile attempt to evaluate designs on the basis of opinion. Specifications drawn up in advance by the DMA provided a yardstick by which officers in the DMA could measure with comparative objectivity the designs and, ultimately, the aircraft produced by the BAP in response to those specifications. Interestingly enough, this decision to measure actual performance against specifications by a proof test was a reversion to the practice begun as far back as 1914 in the first competition held in the United States to procure military aircraft. In the second place, the plan establishing an organization to secure a continuing flow of technical and tactical information from the fighting front marked a decided swing away from earlier reliance on "qualified" individuals with "experience" supposedly sufficient to be used as a basis for making technical decisions.

The end of the war came so soon after the new agreement between DMA and BAP that one cannot really pass final judgment on the success or failure of the arrangement. Nevertheless, changes in organization thereafter were few in number and limited in scope; this may be some indication of success in the newly devised functional assignments.[22] After months of trial and error the successive authorities responsible for tactical aircraft had at last hammered out an apparently effective organization for making decisions. The essential novelty of that organization lay in basing decisions upon

21. Draft of BAP-DMA agreement, Memo, Col. T. H. Bane to Chief of Staff, 1 Aug. 1918, NA, BAP Exec. Box 81, 402.1 Technical Data. See also Copy of agreement sent to Chief of Staff, 1 Aug. 1918, signed by Maj. Gen. W. L. Kenly, Director of Military Aeronautics; Col. H. H. Arnold, Asst. DMA; J. D. Ryan, Director of Aircraft Production; and W. C. Potter, Asst. Director of Aircraft Production, NA, BAP, Exec. Box 6, 026.4 Scientific Research Department.

22. As evidence of the relatively stable organization after 1 Aug. 1918, see, for example, "Organization Outline of Division of Military Aeronautics," 20 Sept. 1918, AAF Archive, M1063-1.

information rather than upon opinion; and gathering information, whether tactical or technical, required organization as much as did decision-making.

Before turning to the problem of organization for information, it may be of some value to consider British experience in developing air materiel as a basis for comparison with the achievement of the United States.[23] A full year passed from the time the Bolling Mission sent its famous report to the date of the working agreement between the Bureau of Aircraft Production and the Division of Military Aeronautics. During that year the program for aviation floundered ineffectively, production lagged, and only a trickle of aircraft reached the fighting front, thereby delaying victory and sharply curbing the pace of doctrinal evolution. It may well be that the year of trial and error was unavoidable. It may also be argued that the time so spent was less than half the time spent by the British in securing the same results. The British had already struggled through nearly two and a half years of organizational chaos before achieving an effective administrative mechanism. Here was an obvious invitation to study British methods if not to emulate them.

Military attachés filed routine reports on the various agencies dealing with aviation in the United Kingdom, and there were interested observers who returned an occasional comment on the workings of the British system for translating ideas into weapons. Nevertheless, few if any of these reports and comments seem to have been given serious analysis in the United States.[24] Organizations for solving problems of air materiel in this country merely grew; they were not planned in any long-range sense of that word.

By the middle of 1917 the British organization for aviation had been forged by the necessities of war to the point where its functions and grouping of functions could be studied with profit. British authorities had come to realize the importance and the relationship of organization for information and organization for decision. Of course, the precise form of any single agency could scarcely be copied exactly since each evolved to suit the peculiarities in the administrative structure of the kingdom. Earlier in the same year

23. J. M. Spaight, *The Beginnings of Organized Air Power* (London, Longmans, Green, 1927), gives a comparative view of French, German, British, and U.S. air organization. Although based almost entirely on secondary sources and rather sketchy, this study presents one of the few comparative analyses available.

24. A careful search of BAP Executive Office files and DMA files in the National Archives failed to reveal more than the most superficial interest in British organization and procedures. The whole field of organizational planning and management control in the United States was in its infancy during the period under discussion.

the British War Cabinet had reshuffled the Air Board to include representatives from the air arms of the army and navy as well as from the ministries for munitions. Here, within a single board directly responsible to the powerful War Cabinet, aerial doctrine, materiel, and programs of production were brought into balance. The Air Board formulated over-all air policy, doctrine, and objectives, then considered the detailed programs submitted by the War Office and Admiralty after their respective aerial policies had been "concerted with" the board itself. Above all, the Air Board was concerned with "selecting and being responsible for" designs for aircraft, engines, and accessories. In short, the Air Board was the sole agency for critical decision. No comparable agency existed for aviation in the United States throughout the period of World War I.

The British Air Board did not depend upon "qualified" and "experienced" individuals for its decisions as did the Joint Army-Navy Technical Board in the United States. To support the function of decision-making by the Air Board, there were a number of subsidiary offices including an administrative Secretariat, a Director of Requisitions and Statistics, an Inventions Committee, a Central Air-Intelligence Division, and a Technical Department staffed with technically proficient officers of both army and navy assigned to carry out proof testing, to keep in touch with scientific advances, and to advise the Air Board about decisions on design.[25] The British had succeeded in concentrating in a single high and powerful echelon the formulation of aerial doctrine and the selection of aircraft to implement that doctrine. The British system relied not upon omniscient individuals but upon an organization supplying facts.

When Colonel Bolling reached Europe the British Air Board constituted a body of experience, at once organizational and procedural, which might well have served as an example to save the United States a year of frustration and a generation of dispute. The opportunity went unexploited, and the early gropings of the British were repeated on this side of the Atlantic.

25. Discussion of Air Board based largely on photostat copy of British air arm administrative manual, *ca.* Dec. 1917. Changes in organizational nomenclature appeared with the formation of the RAF, but functional groupings remained substantially the same. In this regard, see Capt. R. S. Rainsford to Exec. DMA, 24 July 1918, NA, BAP Exec. Box 6, 026.4 A.S., AEF. See, for general discussion, C. G. Grey, *A History of the Air Ministry* (London, G. Allen and Unwin, 1940), chaps. iii–vi; W. A. Raleigh, *The War in the Air* (Oxford, Clarendon Press, 1922), Vol. *1 passim;* and H. A. Jones, *The War in the Air* (5 vols., *2–6;* Oxford, Clarendon Press, 1928–37), Vol. *3,* chap. iv, Vol. *6,* chap. i.

## Chapter V.  Organization for Information

To UNDERSTAND JUST how the organization for securing information evolved it will be necessary to retrace a few steps and refer to the period just after the Bolling Mission's formal report of July 1917. At that time those with authority over the program for aviation only dimly and imperfectly perceived that the pace of changes in design of aircraft at the front required an organization supplying a constant flow of technical and tactical information as well as an organization to make effective decisions based on that information.

The gradual evolution of official opinion regarding the problem of exchange of information is reflected in the mechanisms set up to secure that end. The Bolling Mission, essentially a single-purpose and *ad hoc* agency, was ill equipped to perform a continuing function. It suffered from limitations and deficiencies essentially the same as those of the parties of foreign officers initially sent to the United States. This exchange of missions, which took place in the months following April 1917, failed to establish an effective two-way flow of technical information between the United States and the Allies. But in one respect there was accomplishment. Early reliance upon supposedly omniscient individuals in permanent missions gave way to a concept of revolving assignments in which members of a mission were periodically replaced by officers from the front "with the latest experience." [1] The realization that experienced individuals, no matter how well traveled and

1. For a specific example of an official expression of policy on utilization of "experienced" officers late in the war, see Report of Chief Signal Officer, *Annual Reports of the War Department*, 1918, *1*, 1074. That the whole problem of information as an essential factor in military planning and operations had been seriously neglected by the War Department in general and not just with reference to the aviation program is indicated by P. C. March in *The Nation at War* (Garden City, Doubleday, Doran, 1932), p. 226. On becoming Chief of Staff, March found Military Intelligence a "minor appendage" in the General Staff War Plans Division. In April 1917 there were two officers and two clerks assigned to Intelligence. By the end of the war, there were 282 officers, 29 enlisted men, and 948 civilians. Throughout this chapter and the study as a whole, the term "information" has been used rather than "intelligence" as being less endowed with special connotations.

trained, were never a satisfactory substitute for a continuing flow of information from many sources was to penetrate official circles only slowly and incompletely throughout the war.

The Balfour Mission to Washington, including the original group of British officers sent to the United States in 1917, was, like the Bolling Mission, unprepared to operate on a permanent basis. Nevertheless, by the end of September 1917 the British mission had expanded, notably in creating a technical branch for the express purpose of handling interchanges of information.[2] The British, with some three years of wartime experience behind them, delayed approximately six months before establishing an effective organization to carry out this function in the United States.

In France Maj. E. S. Gorrell, one of the army representatives left behind when the Bolling Mission broke up, recognized the importance of organization for information by the time he had been abroad four months. It was logical that Major Gorrell should become interested in the problem as early as he did. Not only was he an aeronautical engineer, aware of the influence upon design of changing tactical requirements at the front, but as a member of the Bolling Mission he had also been in an exceptional position to see the important relationships of aerial doctrine, tactical application, change in design, and achievements in production. During the summer of 1917 Major Gorrell established a Technical Section in the Air Service, AEF, to carry on the functions initially performed by the Bolling Mission.[3] The plan was probably sound, but it proved difficult to carry out. During August 1917, because of the acute shortage of personnel in the advance party of the AEF, the Technical Section consisted of but one officer, Gorrell himself. No one appreciated the ineffectiveness of such an understaffed organization more than Major Gorrell who was at the same time serving as a member of the General Purchasing Board, an examining board, and a supply department. In addition, he was designated aviation officer on the Line of Communication, AEF. This staggering burden induced the major to write privately, and out of channels, to a friend on the staff in Washington begging for a stronger agency for liaison. A large staff was essential, he contended, because one had to go out and press Allied agencies for

2. R. M. McFarland, "British War Mission to the United States," 21 July 1919, NA, BAP Hist. Box 10, 336.91 British War Mission.

3. Final Report of Chief, Air Service, AEF, draft copy, chap. xii, p. 101, 1919, NA, WWI Orgn. Records, A.S. Hist. Records Box 1.

information rather than wait for gratuitous releases. To fail in this function, the major implied, was to allow the United States to waste precious time in developing technical equipment long since perfected by the Allies.[4]

Major Gorrell's comments are revealing. They show that he saw the need for a system of information in place of a roving mission even before Bolling's group had ceased to operate. Moreover, they suggest that Gorrell had an early insight into the importance of aggressive rather than passive information-gathering. Experience was to show how slowly this idea gained headway in aviation circles. Despite the example of the British mission in Washington and Gorrell's Technical Section of the Air Service, AEF, aviation authorities in the United States continued to place their reliance upon individuals and missions rather than upon organizations as a source of data.

The Waldon Mission was a typical result of the policy which relied on roving groups for technical information. During November 1917 Col. S. D. Waldon, who had come into the aviation program as an expert on production by way of the NACA, was sent to France to transact some business regarding the facilities to be provided by the Allies for manufacturing aircraft. In addition to this assignment, the colonel was instructed to gather information on developments in aircraft, especially German types.[5] It seems not to have occurred to the authorities assigning Colonel Waldon to his task that information gathered by an individual as a secondary function during a hurried visit must almost necessarily consist of subjective opinion based on evidence casually and haphazardly acquired.

Colonel Waldon's trip to France brought to light one of the factors which contributed to the protracted delays in the development of designs in the United States. The colonel discovered that all cabled messages to Washington passed through a single wire office at AEF headquarters which constituted a serious bottleneck.

4. Copy, Maj. E. S. Gorrell to Maj. [W. L.] Souther (initials uncertain), 4 Sept. 1917, NA, BAP Hist. Box 10, 334.8 Bolling Mission.

5. Two draft studies, one by Capt. J. L. Ingoldsby and the other by R. M. McFarland, both entitled "History of Waldon Mission," the latter version dated 30 Oct. 1919, constitute the basis for this paragraph. NA, BAP Hist. Box 10, 334.8 Waldon Mission. See also, Photostat of signed copy, "Final Report of Special Mission Overseas, Bureau of Aircraft Production," 1 Nov. 1918, hereafter cited as Lockhart Report. This copy bears marginal criticisms by Col. E. S. Gorrell, written 19 Aug. 1919, questioning the validity of superficial joy-riding missions: "In the next war," wrote Gorrell, "avoid missions of this kind." NA, BAP Hist. Box 10, 334.8 Lockhart Mission.

It required an average of three to five days to clear a message through this cable system, and on occasion letters sent by steamer reached Washington before messages sent by wire.[6] Cable jams were not the only cause of difficulty in transmitting technical information from the front. When Colonel Bolling had first sailed for Europe, his broad credentials gave the mission a quasi-diplomatic standing. This status, along with the presence of the colonel's brother-in-law William Phillips, who occupied a post in Washington as Assistant Secretary of State, gave support to Bolling's use of the State Department's transatlantic line in reporting some of the mission's findings to Washington. Cables sent through diplomatic channels were apparently not cleared with GHQ, AEF, and the possibility of contradictory reports reaching the United States was greatly increased.[7] Any attempt to evaluate the feasibility of personal missions to the front as a source of information must take into account the multiplicity of channels and crowded wires, not to mention the confusions attending garbled language in cabled messages, which impaired the flow of data from Europe to the United States. During the summer and fall of 1917 all these factors must have encouraged the practice of relying upon individual missions for data about designs.

Instructions drafted for a subsequent trip by Waldon reflect the prevailing attitude in Washington. They directed the colonel to inform the Chief of the Air Service in the Zone of Advance of the "grave necessity" for having officers traveling continually back and forth every 30 or 60 days to bring a "personal point of view" to Washington. Personal impressions of the front were important to those confined to Washington offices, but personal impressions were here used in lieu of rather than as a supplement to detailed technical information compiled systematically from a large number of sources. While the importance of Colonel Waldon's "personal point of view" cannot be denied, it was manifestly impossible for him to carry out his assigned task. This wholesale order involving thousands of details included securing "the latest information on planes, engines, and equipment." The colonel's direc-

6. Transcript of testimony by Col. S. D. Waldon before special meeting of the Aircraft Board, 8 Feb. 1918, NA, BAP Exec. Box 65, 334.8 Aircraft Board.

7. Testimony of Maj. B. D. Foulois, *War Expenditures,* House Hearings, 66 Cong. 1 Sess., Serial 2, Pt. 6, 6 Aug. 1919, pp. 393–4. Even a casual reading of AEF cables pertaining to aviation on file in the National Archives reveals the difficulties encountered because of poor communications. Decoding or transmission errors are frequent, and messages are occasionally in such cryptic cablese as to defy comprehension. Marginal queries and requests for repeats indicate that the recipients in 1917 and 1918 suffered the same confusion as a contemporary reader.

tive assigned him functions sufficient to busy a whole staff of technical experts.

In addition to his other duties, Colonel Waldon was to find out how the Technical Section, AEF, was organized and what methods the section employed in securing technical data.[8] These instructions indicate that as late as May 1918 aviation officials of the Signal Corps in Washington were ignorant of the structure and operation of the Technical Section, AEF. Nor were they utilizing its services fully; technical data was not reaching the United States in the quantity or at the speed desired. Nevertheless, while making only the most perfunctory efforts to improve the organizational channels of liaison, the authorities in Washington continued to rely upon the services of messengers or missions.

In April 1918 André Tardieu, French High Commissioner to the United States, suggested to the Secretary of War that it might be possible to save three months in the time spent planning for production of the SPAD fighter aircraft by sending a mission to study French industrial techniques for that particular type.[9] One month later the head of the mission detailed to secure the information proposed by the French commissioner received his instructions. After a further delay of two weeks still another set of instructions was drafted for a more qualified engineer appointed to replace the original head of the mission. Thus at least half the time supposed to be saved by sending a mission was lost even before the mission departed for France. Yet in spite of the administrative delays encountered in setting up the cumbersome machinery of a mission, the directive presented to the chief of the group enjoined him to work with all possible speed: "The important thing that you must keep in mind is that we want the very latest and most up-to-date design . . . We want to put this design into quantity production in this country at the earliest possible moment." The author of this directive added, almost as an afterthought, that the desired information could be secured either personally or through the Technical Section, AEF, whichever seemed best.

The instructions for the chief of the SPAD mission reflect something of the lack of clearly defined policy regarding the selec-

8. Lt. Col. L. S. Horner, Exec. Dept., Equip. Div., to Col. S. D. Waldon, 20 May 1918, NA, BAP Hist. Box 10, 334.8 Waldon Mission. For pertinent comment on the importance of "bulk" information over individual "critical" items, see G. S. Pettee, *The Future of American Secret Intelligence* (Washington, Infantry Journal Press, 1946), pp. 34–7.

9. André Tardieu to Secretary of War, 18 April 1918, NA, BAP Exec. Box 64, 334.8 Bechereau Mission.

tion of designs for aircraft which still prevailed in the early spring of 1918. Although the mission had originally been proposed as a means of transporting skills and techniques in production from France to the United States, the instructions as finally drafted made the group responsible for securing information about production and data on performance, including "the opinions of Air Service test pilots." [10] Moreover, implicit in the instructions was the assumption that the mission would reach decisions as to types of aircraft. Like the previous Bolling Mission, the SPAD mission was given a variety of tasks with a vague area of implied powers to make decisions upon the matter of design. Here in its simplest form was the classic contest between production and operations as discussed in the preceding chapter. The SPAD mission represented the point of view of production. The aviation officers of the AEF who were responsible for the use of aircraft in operations against the enemy were little inclined to allow the critical decisions of design to be made by representatives of production in the United States. Therefore, the Chief of the Air Service, AEF, formed a board to do the same thing.

On the day that Tardieu proposed the SPAD mission to Secretary Baker in Washington, Brig. Gen. B. D. Foulois appointed a board of AEF officers in Paris "to investigate and report upon all new types of airplanes and engines . . . being considered by the French and English . . ." The board spent five weeks studying aircraft in England and France, then submitted a report. Looking back on the event after the war, the president of the board was inclined to blame the liaison "mess" on the authorities in Washington. [11] Nevertheless, the existence of a special body to secure information on designs reveals a confusion of functions within the AEF itself, where a Technical Section already existed to execute tasks similar to those assigned to the board. If the board had been assigned the specific task of reaching decisions on the basis of data compiled by the Technical Section there would have been no duplication of function, but the board's directive did not so limit its scope. The liaison "mess" was universal throughout all echelons at the battlefront as well as on the home front.

10. Mission Instructions to Captain Page, 15 May 1918, and to A. V. Verville, 28 May 1918, NA, BAP Exec. Box 64, 334.8 Bechereau Mission.
11. "Formulation and Distribution of Programs," Sept. 1919, NA, BAP Hist. Box 9, 334.8 Overseas Missions; and notes on conversation with Col. T. F. Dodd by R. M. McFarland, 8 April 1919, NA, BAP Hist. Box 20, 452.1 Airplanes, General. For caustic comments on Pershing's aircraft boards, see March, *Nation at War*, pp. 283–4.

After the Bureau of Aircraft Production (BAP) and the Division of Military Aeronautics (DMA) had reached their working agreement during the summer of 1918, the chief of production J. D. Ryan wrote to General Pershing in France explaining the situation as he saw it. He noted that President Wilson's executive order of May 1918 had, broadly speaking, assigned all operational functions to the DMA and all functions of supply to the BAP in the United States. On the other hand, in France both operations and supply fell within the province of the Chief of the Air Service who was, so Ryan believed, more concerned with the military use of aircraft than with supply.

With General Pershing's approval, Ryan therefore directed Col. P. L. Spaulding to establish an agency in France to supply the BAP with "continuous information" concerning types and numbers of aircraft needed and to secure reports on performance achieved by models already in the field. In addition, the new agency was to keep in close touch with the progress of research in Europe and in the United Kingdom in order to provide information for the National Research Council in the United States. To carry out these many duties Colonel Spaulding was given an empire builder's dream, a free hand in requisitioning personnel and equipment to organize an office.

Spaulding's directive is noteworthy for several reasons. For one thing, it proposed the establishment of an organization rather than a temporary mission. This arrangement indicated a better appreciation of the problem of information than there had been hitherto. The whole concept of the plan to send "continuous information" marked a substantial step beyond the thinking embodied in directives to previous missions. On the negative side, Spaulding's directive ignored the question of channels for routing information back to the United States. Moreover, it failed to instruct the new agency to coordinate its activities with existing agencies for processing information, such as the Technical Section of the AEF.[12]

At about the same time that Colonel Spaulding was busy establishing his information-gathering organization, another mission from the BAP was operating in Europe. In response to a request from the AEF, the Assistant Director of Aircraft Production authorized a mission to France under the leadership of a dollar-a-

12. R. M. McFarland, "History of P. L. Spaulding's Mission Overseas," 4 Nov. 1919, NA, BAP Hist. Box 10, 334.8 Spaulding Mission. For an insight into the BAP idea of the Spaulding project, see Lt. Col. L. S. Horner to Brig. Gen. M. M. Patrick, Chief, Air Service, AEF, 22 July 1918, NA, BAP Exec. Box 81, 402.1 Technical Data.

year expert on production, Henry Lockhart. The directive for the
Lockhart Mission was reminiscent of Colonel Bolling's instructions.
Despite the months which had passed and the experience which had
been encountered in dealing with numerous missions overseas,
Lockhart's directive assigned duties sufficient for a regiment.

The mission was expected to obtain from the Allies information
on their requirements in types of aircraft for a two-year period.
In addition to that patently impossible task, the mission was to
study industrial methods, discuss aviation matters with foreign air
ministers, review the status of foreign experimental work, and interview officers at the front to gather information as to performance of aircraft in combat. Even more surprising than the variety
of functions assigned to the mission was the extent to which it
received authority to participate in decisions on design. Lockhart
was specifically instructed to "come to a definite understanding
with the Commander of our Air Forces as to which types should be
produced . . ." [13]

Among the many functions assigned by Lockhart's directive,
perhaps none was so significant as the instructions to establish a
bureau of information for regular and constant liaison between
France and the United States. Apparently there had been no coordination within the BAP in drafting instructions successively
for the Lockhart and Spaulding missions. Something of the motives
which led the BAP to spawn information agencies as it did can be
detected in Lockhart's final report. There existed among the European experts on aviation, the mission found, "an extreme diversity
of opinion" as to the relative merits of the different types of aircraft along the front. Far from reaching any clear and final decisions as to types, the mission discovered that it was all but impossible to get any two pilots or even two engineers to agree fully
upon types of aircraft for production.

Back in the United States officials of the BAP who were responsible for production in quantity were frantic for authoritative
information upon which they could plan with assurance. Each mission they sent to Europe labored, apparently, in the forlorn hope
that it could secure that vital information. Like all the others
the Lockhart Mission brought recommendations for better methods
of handling information rather than conclusive information itself. The realization dawned only very gradually in aviation circles
that there would never be any conclusive information on designs

13. Lockhart Report, 1 Nov. 1918, and R. M. McFarland, "History of Lockhart
Mission," 30 Oct. 1919, NA, BAP Hist. Box 10, 334.8 Lockhart Mission.

for aircraft. The design of air weapons was in constant flux, and only a continuous flow of information, both tactical and technical, could keep the Air Service, AEF, superior to the enemy.

Colonel Spaulding's service duplicated the work of existing agencies and curtailed, to a certain extent, the substantial beginnings already made in establishing formal liaison with the air ministries and technical sections of the several Allies. The absorption of liaison functions by the new agency would not in itself have been a negative step, but after confusing the pattern of information agencies and impairing their effectiveness Colonel Spaulding was abruptly ordered to Washington. The information channel which he had begun to organize dried up, and the BAP was about as frantic for information as it had ever been.[14]

The Lockhart report, completed on the first of November 1918, came too late to have any real influence upon the war in progress. In the confusion attending the closing days of the war it is unlikely that the report received much attention, although it contained a comparatively advanced appreciation of the problem. Lockhart specifically recommended that a permanent organization to collect information be established in Europe and that all future representatives and missions sent to Europe from the various agencies in the United States be routed through this permanent organization. In making detailed suggestions regarding the composition of the new office, the Lockhart report differed significantly from the reports of previous missions. Lockhart believed that the new office should have a formal table of organization with branches properly staffed to perform the function of liaison with European agencies for science and research, engineering, production, and inspection.[15] In addition, he favored a staff of trained specialists operating in functional units. There was implicit in this proposal the concept of information as a commodity to be handled in the same manner as any other commodity, objectively and systematically.

Officials of the BAP continued to send repeated missions to France long after the Technical Section, AEF, had been organized. Their faith in missions suggests either a lack of confidence in the work of the Technical Section or a failure to receive the information required. Probably both factors were present. Compared to the poorly staffed missions of the BAP, the Technical Section, AEF, was well provided with personnel, although when it had been established during the summer of 1917 it had consisted of a single

14. Lockhart Report.
15. Lockhart Report.

overworked officer who enlarged the staff by hiring some citizens of the United States who happened to be in Paris at the time. By November 1917 the Technical Section had a staff of 50, including 19 officers, and in November of the following year there were 457 employees in the section including 126 officers. The structure of the Technical Section expanded with its increasing personnel, and as early as November 1917, a full year before the Lockhart report appeared, there were already eight separate functional divisions, with specialists in aircraft, instruments and equipment, armament, engines, photography, meteorology, information and statistics, as well as miscellaneous mechanical problems. Nevertheless, it was not until the final months of the war that the Technical Section added a History and Research Division to gather tactical information to help designers improve aircraft. It may be recalled that the Civil War was ending before the Ordnance Department recognized the importance of this kind of information for developing weapons; the Air Service repeated the pattern and appreciated this need only at the end of World War I.

A member of the staff summed up the functions of the Technical Section at the time of the Armistice in a single sentence: "to improve the over-all efficiency of the Air Service, AEF, by making recommendations regarding the best designs for aircraft and equipment." Specifically, the section was to conduct studies of all Allied and other equipment as well as all patents and inventions in order to submit reports recommending decisions on technical questions to the Chief of the Air Service. Some conception of the responsibilities held by the section (or believed by its personnel to be so held) is indicated by its directive that all aviation materiel be "approved by the Technical Section before being officially adopted." [16] It is noteworthy that the list of duties for the Technical Section compiled by its staff at the end of the war did not include any mention of responsibility for sending information to the United States, even though the Chief of the Air Service considered that to be one of the agency's primary responsibilities.

The Technical Section, AEF, was well equipped to gather in-

16. "Progress of Air Service Activities as of 11 Nov. 1918," A.S. Exec. Sect., 30 Nov. 1918, National War College Library, UG576.3; and Final Report of Chief, Air Service, AEF, draft copy, chap. xii, pp. 101–7, 1919, NA, WWI Orgn. Records, A.S. Hist. Records Box 1. See also A.S., AEF, Memo No. 75, 7 Oct. 1918, quoted in H. A. Toulmin Jr., *Air Service, American Expeditionary Force, 1918* (New York, D. Van Nostrand, 1927), pp. 307–8. Toulmin (p. 310) gives an organization chart, without authenticating signature, in which the Technical Section in the U.S.A. is shown as a branch of the Technical Section, AEF.

formation. In addition to its specialized engineers, the section
maintained a subbranch for liaison with London as well as liaison
officers in the technical section of the French air arm and a single
liaison officer in Italy. There was of course a considerable difference
between having the machinery for liaison and actually making it
work. All the available evidence suggests that most of the members
of the staff in the Technical Section were occupied in testing aircraft and equipment before preparing decisions for the Chief of
the Air Service. The mechanics of gathering and sending information to the United States were given comparatively little attention,
and only belatedly was an organization, the Technical Data Division, set up within the section to process information.

The Technical Section stumbled over the mechanics of handling
information. The section was adequately staffed, and the functional structure of the organization prevented decisions on engineering from being thrust upon unqualified personnel as was the
case in small, short-lived missions. But the emphasis on making
decisions led the Technical Section to neglect or slight its responsibilities for processing information. It was confronted with a
multitude of immediate and pressing problems of a technical
character which it solved expediently. There was little time and
probably equally little inclination to establish a careful system of
reporting throughout the facilities of the section, which included
offices in Paris and in the testing center at Orly airfield as well as
the previously mentioned branch offices and liaison officers. Questions arising in the Zone of Advance were acted upon by representatives of the Technical Section who traveled back and forth
between Paris, the front, and the other installations of the organization.[17]

Individual officers moving to and fro were a poor substitute for
a system which required objective reporting and scientific or statis-

17. Final Report of Chief, Air Service, AEF, draft copy, pp. 101–7, 1919, NA,
WWI Orgn. Records, A.S. Hist. Records Box 1. For an indication of the liaison
organization planned for the AEF in the United Kingdom, France, and Italy, see
Copy of unsigned memo, Asst. Chief of Staff for the Adjutant General, 22 July 1918,
NA, BAP Exec. Box 81, 402.1 Technical Data. See also Toulmin, *Air Service, American Expeditionary Force,* who emphasizes the lack of leadership and effective staff
coordination which made "practically a complete failure" of the Air Service, AEF,
until the appointment of Brig. Gen. M. M. Patrick, an officer from the Corps of
Engineers, as chief in May 1918 (pp. 74–5). The Technical Section's role in making
decisions for the Air Service chief, as contrasted with the role of disseminating information, takes on particular importance when one realizes that as late as June 1918
all technical decisions regarding modifications in aircraft were sent all the way up
for the chief's approval for want of an adequate subordinate agency to take care of
the matter. See *ibid.,* pp. 87–8.

tical techniques for fact gathering. During the last few months of the war the Technical Section finally did create a system for statistical reporting up and down the front, but the data so acquired was sadly limited in scope.[18] The difficulty encountered in "educating" the many agencies concerned with the aviation program in the importance of using the channels of information established by the Technical Section was more serious. Whether it was the agencies with their constantly changing personnel which caused the trouble or the failure of the section itself to take sufficiently aggressive action one cannot say. In any event at no time during the war did information flow smoothly and continuously from the Technical Section, AEF, to the several agencies which required tactical and technical information in the United States.

Soon after Major Gorrell created the Technical Section, AEF, it was discovered that agencies in the United States such as the NACA and the Aircraft Board were requesting information directly from French officials who had in many instances already provided identical data to the Technical Section. Duplicate requests of this nature naturally irked the French authorities and gave substance to Lloyd George's subsequent scathing comments on the capabilities of organizers in the United States.[19] Aviation officers in Europe were inclined to blame the failures of the system of liaison upon the "absolute lack of support" received from Washington. There was some justice to this contention. It was not so much a matter of official unwillingness to be cooperative as it was a general lack of understanding by those in the ever-changing organizations in the United States of the proper channels and agencies with which to deal.[20] A single instance will illustrate the problem adequately. As late as June 1918 the BAP executive office found it necessary to point out to members of the staff that the office of the Technical Section in London was a subbranch of the Technical Section, AEF, in France and all dealings with the Lon-

18. After June 1918 statistical reports were regularly compiled on such details as flying hours, aircraft available, wastage, personnel strength, spares, etc. See "Progress Reports of Air Service Activities, AEF," A.S. Exec. Sect., 1918, National War College Library, UG576.3.

19. Extract from letter by Lt. Col. E. S. Gorrell, 29 Oct. 1917, NA, BAP Exec. Box 81, 402.1 Technical Data; and David Lloyd George, *War Memoirs* (6 vols.; Boston, Little, Brown, 1933–37), *5,* 451: "It is one of the inexplicable paradoxes of history, that the greatest machine-producing nation on earth failed to turn out the mechanism of war after 18 months of sweating and toiling and hustling. The men placed in charge of the organization of the country for this purpose all seemed to hustle each other—but never the job."

20. S. G. H. (author otherwise unidentified) to Henry Lockhart, 16 May 1918, NA, BAP Exec. Box 81, 402.1 Technical Data.

don office should be routed through Paris.[21] One need not attempt to weigh the pros and cons of the procedure itself, but the evidence shows that months after the Technical Section, AEF, had been established as an operating agency officials in the United States, whose activities required almost daily cooperation with the section, apparently remained ill informed as to its organizational structure and operating procedures.

Had the Technical Section, AEF, devoted more time and energy to "educating" officials in Washington, this difficulty might have been alleviated but not necessarily resolved. An all-embracing solution to the problem of liaison probably lay beyond the scope of any administrative measures the Technical Section was capable of taking, since the sources and channels of information were threaded throughout the entire Air Service, AEF.[22] Moreover, the Technical Section shared the "function" of gathering and disseminating information with another agency, the Information Section of the Air Service Headquarters at Tours, to which it was in many ways related.

The Information Section, AEF, was an outgrowth of the office of intelligence which had been organized for the Air Service Training Department in Paris during 1917. At the end of the war the Chief of the Air Service described the Information Section as the central agency for "technical, military and aeronautical data." [23] This inclusion of technical data in the functions of the Information Section appears to have been an error, for the conception of function held by those actually in the Information Section during the war was limited to information on training, operations, tactics, and the like.[24] Just where the line of division came between the Technical Section and the Information Section, apart from their geographical separation, is hard to determine: the relationship between tactical and technical information is such as to defy separation.

21. Interoffice memo, BAP Exec. Officer to Maj. [H. W.] Jones (initials uncertain), 12 June 1918, NA, BAP Exec. Box 81, 402.1 Technical Data.

22. Air Service, AEF, Organization Chart for 6 Sept. 1918 shows the complexity of the sources of information, which include air units with armies in the field; at four headquarters, Chaumont, Paris, Tours, and London; foreign liaison offices; the Coordination Staff; and others, for the most part in echelons above the Technical Section. NA, BAP Exec. Box 6, 026.4 BAP.

23. Final Report of Chief, Air Service, AEF, draft copy, p. 51, 1919, NA, WWI Orgn. Records, A.S. Hist. Records Box 1.

24. See, for example, functions of Information Section listed in "Progress of Air Service Activities as of 11 Nov. 1918," A.S. Exec. Sect., 30 Nov. 1918, National War College Library, UG576.3.

*Schematic Diagram Illustrating the Complexity of the System
for Transmitting Technical Information from Europe
to the United States During 1918*

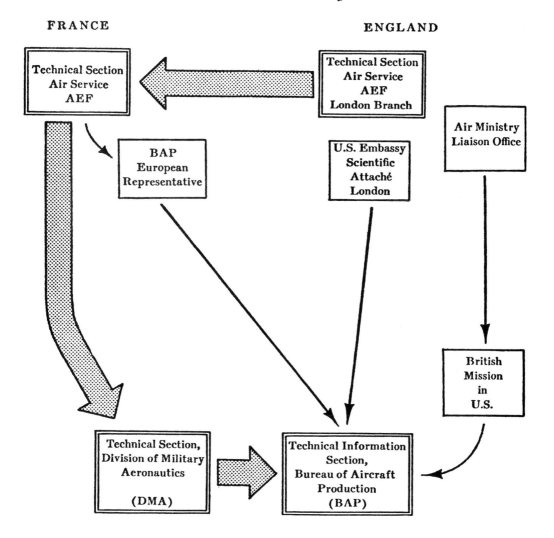

FRANCE

ENGLAND

Technical Section
Air Service
AEF

Technical Section
Air Service
AEF
London Branch

Air Ministry
Liaison Office

BAP
European
Representative

U.S. Embassy
Scientific
Attaché
London

British
Mission
in
U.S.

Technical Section,
Division of Military
Aeronautics

(DMA)

Technical Information
Section,
Bureau of Aircraft
Production

(BAP)

UNITED STATES

*Broad lines indicate normal channels of command; other lines indicate expedient
channels established by Bureau of Aircraft Production officials. Even on this
schematic and highly simplified diagram, it is evident that multiple sources of design
information reaching the BAP made conflicting recommendations almost inevitable.*

The split between the Technical and Information sections, AEF, is more significant than may at first appear. To begin with, the duplicate or overlapping functions of the two agencies had a considerable influence upon the operation of the whole program of production during World War I. Even more important, the split between tactical and technical information, though never complete, was to establish a precedent which drove a wedge between the two for some 20 years following the war, tending to disassociate the air weapon as a technical achievement from its application in combat. Each of these influences upon the air weapon exerted by the structure of the organization for liaison must be studied in detail. Chronologically the influence on production came first.

Inadequate tactical and technical information was the root cause of the delays which held back the program of aircraft production in the United States. The Technical Section's failure to advertise its functions and procedures aggressively, coupled with the want of clearly defined areas of interest between the two agencies for technical data, resulted in an inadequate flow of information to the United States. During the summer of 1918 officials handling production became so concerned with the breakdown of the system for information from Europe that they went so far as to establish their own system within the AEF. This unsuccessful venture, which has already been described in the account of the Spaulding Mission, was in effect a criticism of the failure of the AEF to supply the information necessary for production of aircraft.

When the BAP created a new Technical Information Section in Washington it was an implicit admission that the fault was not entirely confined to the overseas Air Service but lay also in the absence of an adequate agency for processing information in the United States. To improve the flow of technical data, this new section in the BAP secured approval from the Chief of the Air Service, AEF, of a plan to have all information, which was normally sent to the DMA, duplicated and sent to the BAP as well.[25] Thus, by August 1918 the organization for passing information from the fighting front to the point of production consisted of two parallel systems. One was the normal channel of command from the Technical Section, AEF, across the Atlantic to the Technical Section, DMA, and thence, in the form of directives, to the BAP for certain types of aircraft. The second channel of information was that ex-

25. Lockhart to Chief, Air Service, AEF, 14 August 1918, and Copy, Memo by Chief, Air Service, AEF, 15 August 1918, NA, BAP Exec. Box 81, 402.1 Technical Data.

tending from the BAP representative directly to the BAP Technical Information Section. This parallel pattern was somewhat complicated by the order issued by the Chief of the Air Service providing that duplicates of all reports sent to the DMA be sent direct to the BAP. Despite these irregularities, by the end of the summer of 1918 the director of the Technical Information Section of the BAP was able to report that the flow of information was accelerating and that liaison was generally excellent. The illusion was short-lived.[26]

Little more than a week after the Director of Information for BAP had pictured the problem of liaison in glowing terms, a new storm broke out. The Technical Section (DMA) protested, not without some justification, that the newly created Technical Information Section of BAP usurped and duplicated functions properly belonging to the DMA. A suggestion by officials of the DMA that the new section be abolished appears to have had little effect since the Technical Information Section continued to exist. As long as the DMA failed to improve the mechanics of its organization for disseminating and processing information, it could expect the BAP to operate rival and parallel agencies. Officials of the BAP insisted, but failed to support their contention with examples, that processing by the DMA delayed information by as much as four weeks, thus making it "stale" before it reached the production officials who needed it.[27]

The arrangement certainly did not simplify the handling of information. An impartial observer might well sympathize with an unhappy British official in Washington who admitted with characteristic understatement that he was "extremely exercised" over the situation regarding technical data. Where, he asked, should information be sent, and what guarantee had he that sending it to one person would ensure its proper circulation?[28] His question cut to the heart of the problem. The duplicate organizations set

26. Director of Tech. Info. Sect. to Asst. Director, Aircraft Production, 9 Sept. 1918, NA, BAP Exec. Box 81, 402.1 Technical Data. W. C. Sabine, the Technical Information Section director, was put in charge of the NACA Office of Aeronautical Intelligence in September 1918. This aviation information agency, a further complication in the pattern, had been established in January 1918 as a "central governmental depository" for aeronautical data. It relied upon the National Research Council (see following chapter) for information from Europe. NACA *Annual Report,* 1918 (Washington, GPO, 1919), pp. 24–5.

27. Asst. Director, Aircraft Production, to Acting Director, 18 Sept. 1918, and Memo, BAP Exec. to Acting Director, 31 Oct. 1918, NA, BAP Exec. Box 81, 402.1 Technical Data.

28. Col. W. Simphill, Special British Mission, to Asst. Director, Aircraft Production, 21 July 1918, NA, BAP Exec. Box 81, 402.1 Technical Data.

up by the BAP were irregular emergency measures, more expedient than effective, designed to solve a problem by direct action. In doing so, they created a whole series of subsidiary problems. For each new information agency established, there were countless existing agencies which would have to be "educated" as to appropriate channels and methods of operation.

Just how difficult the process of orientation could be is illustrated by the relationship of the BAP Technical Information Section with the British Mission in Washington. The staff of this mission built up a working relationship with the Technical Section of the DMA. When the new Technical Information Section appeared on the scene, a whole new set of relationships had to be determined. The problem was not simply one of handing out duplicate information, for the system of liaison was far more complex than that. The reader will remember that the Technical Section, AEF, maintained a branch office in London. The BAP had no such office, so it was compelled to rely more heavily upon the liaison services of the British Mission unless it was willing to wait for information to be sent from the branch in London to the office of the Technical Section, AEF, in Paris and thence to the Technical Section of the DMA in Washington for processing to the BAP.

The situation presented by the system of multiple sources of information was not unlike the complexity of channels confronting the foreign office of any modern state. Diplomatic information from abroad can be gathered either through a minister or ambassador residing in a foreign capital or through the minister of a foreign state accredited to the government seeking information. The analogy is useful because it reveals that the authorities in the United States were not entirely without precedent in struggling to solve their problems of liaison even though they failed to derive much profit from the precedent. Whether officials of the BAP sought information through the DMA or through the British Mission was undoubtedly determined in the final analysis by the success of one or the other source in providing the data desired. Regardless of comparative effectiveness, both channels continued to function and provided duplicating and overlapping as well as conflicting advice.[29]

Failing to receive what the authorities in charge of production considered an adequate flow of technical information through the military channels of the DMA, the director of the Technical In-

29. Some of the difficulties arising from conflicting advice from abroad are discussed below in chapters 7 and 8.

formation Section (BAP) set out to improve direct liaison with the British. In an accord reached with a representative of the British Aviation Mission in Washington, it was decided that all future exchanges of information would be cleared through a single individual in London dealing "on the widest possible lines." All specific requests for information were to be channeled through this person, and the reports of all missions, deputations, or observers in the United Kingdom were to be returned through him. Finally, it was stipulated that all information returned to the United States would be disseminated by a single suitable agency. Presumably the suitable agency was the BAP Technical Information Section.[30] The agreement of the BAP and the British Mission contained several administrative features which promised to help untangle the snarl of agencies for liaison. If it did nothing more than reduce the number of independent and uncoordinated reports returned by missions and observers, the agreement would have been useful. Nevertheless, creating a new agency in London, even if it consisted of only one individual, was after all merely adding one more agency to the many already existing. The agreement was typical of the prevailing pattern and practice. Officials sought to solve problems with more agencies rather than with better agencies, and in creating new agencies the circle of problems widened just so much farther.

How much more difficult the addition of new offices could make the task of acquiring technical information is illustrated by an incident in London. When an official of the Bureau of Standards wrote to the Scientific Attaché at the United States Embassy in London for a certain bit of information, the attaché explained his difficulties in securing what was wanted. The British had set up a Liaison Office in the Air Ministry to handle requests for technical data, and applications for information were channeled through this single agency. The report of the attaché is perhaps the best possible index of the effectiveness of this paper-processing agency. "I succeeded in evading this requirement in some cases; in many other cases it resulted simply in the delay of the information from a fortnight to a month." [31] The vicious circle was complete. Since few, if any, relied upon the established channels of information,

30. "Memo Regarding Technical and Other Information Given to the USA by the Aircraft Production Department of the Ministry of Munitions, Great Britain," 25 July 1918, signed by Dr. Wallace Sabine and Sir Henry Fowler, NA, BAP Exec. Box 81, 402.1 Technical Data.

31. H. A. Bumstead, Scientific Attaché, London, to S. W. Stratton, Bureau of Standards, 30 July 1918, NA, BAP Exec. Box 81, 402.1 Technical Data.

virtually every organization created an agency of its own, increasing the frequency of conflicting reports and reducing the degree of control exercised by the military officials, who were responsible ultimately for the success or failure of the various programs for aviation.

The futility of creating new channels for information was recognized during the fall of 1918, and at least one agency sought to improve the situation by refining existing procedures. The reforms proposed by the Director of Information for the BAP read like a catalogue of errors in the liaison system in operation during the last month of the war. Henceforth, the director suggested, officers detailed abroad for liaison service with the Allies or as members of missions should be chosen from among those "directly responsible for the subsequent use of the information." While this was manifestly impossible in every instance, the director felt it was essential for each officer sent overseas to be "conversant with the subject" of his mission. The director further recommended that the orders of each mission should be "carefully and narrowly defined" by the agency originating the mission and that activities should be confined within the limits of these directives. For a corollary, as if to emphasize the point and take note of very human past experience, the recommendation added that each officer should "occupy himself solely with the specific purpose of the trip and return at the earliest possible moment." Perhaps the most significant advice the director offered was that each returning mission or observer should prepare a written rather than a verbal report and distribute copies to all interested agencies.[32]

The reforms proposed by the official of the BAP reveal continued reliance upon missions or individual observers, even at the end of the war. The emphasis on written reports indicates that officials in the War Department at long last appreciated the slight value of information not reduced to writing. The contradiction inherent in the director's recommendations probably escaped him. If missions were to reduce their findings to report form, why then was a mission necessary at all? Missions, it must be admitted, had certain advantages for special projects over the services rendered by permanent liaison organizations. But for the routine transfers of information, established agencies with functional staffs and well-cultivated official contacts had shown themselves more capable of

32. Memo, Director, Tech. Info. Sect., to Acting Director, Aircraft Prod., 28 Oct. 1918, NA, BAP Exec. Box 81, 402.1 Technical Data.

gathering information than had missions, even if the agencies were deficient in techniques of dissemination.

Officers in the AEF when reviewing the experience of the war reiterated the recommendations of the BAP. Their reports noted that the absence of specific directives and carefully defined instructions had impaired the efficiency of missions, and the practice of assigning personally or professionally unqualified officials had been a constant source of embarrassment. More significantly, criticisms from the AEF went beyond those propounded by the BAP and stressed the importance of agencies for liaison in preference to individual observers and missions.

Looking back on the tangle in liaison, the officers of the AEF who reviewed the whole period of the war emphasized a lesson recognized only after repeated failures: it has been demonstrated, they said, that departments, headquarters, and armies "cannot be depended upon" to furnish information "automatically." Informational agencies could ensure a continuous flow of technical data only by establishing a system of liaison with representatives in every agency for training, operations, or staff duty, each specifically charged with the sole mission of gathering information.[33] On both sides of the Atlantic military officials recognized, at last, that technical information required organizations to process it continually if production was to keep pace with the flux in design required to maintain aircraft superior to those of the enemy. Unfortunately, even these simple fundamentals were not recognized until after the fighting had stopped. Organizations so painfully assembled during the war vanished. Some lessons were salvaged. Many had to be learned over again.[34]

Whether or not the aircraft produced in the United States would be superior, only the competitive test of aerial combat could tell. But before going on to analyze some of the results produced during the course of the war by the disorganized state of the agencies for gathering information and making decisions, one must explore still

33. See Information Section, AEF, in volume entitled "Lessons Learned" in "Air Service, AEF, History," Series A, Vol. *15, ca.* 1919, NA, WWI Orgn. Records, GHQ, AEF Files. See also bibliographical note, below.

34. The Information Section, AEF, anticipated the recommendation of the Lockhart Mission in compiling an extensive history of the Air Service, AEF (see above, n. 33). The "lessons learned," however, were for the most part impressions of the period at the end of the war rather than the product of day-to-day operating experience from 1917 onward. Because of the comparative inaccessibility of these histories after World War I, even the limited value of the "lessons learned" was little exploited.

another area of administration. Along with the several sources of technical data in foreign countries, there were in the United States a number of agencies for engineering and scientific research contributing to the field of aircraft. No study of the aerial weapon would be complete without some consideration of the organization created to utilize these agencies in the national program of aviation.

# Chapter VI. Organization for Research and Development

WHEN THE United States entered World War I the nation was unprepared to meet the demands for aircraft which full-scale participation in a world conflict required. Even though the unpreparedness of the United States made it imperative to count on Europe for immediate production of tactical aircraft, from the very outbreak of war there had been no intention of continued reliance upon foreign states. Pride alone, not to mention strategic necessity, would have compelled the nation, sooner or later, to establish a self-sufficient aircraft industry independent of foreign leadership. A self-sufficient source for aircraft implies the existence of a highly perfected system for converting scientific and technological advances into designs for production. Such a system, in its barest essentials, would require organizations for seeking and exploiting new scientific knowledge and organizations for applying in practice the findings of the scientists. In short, the nation needed agencies for research and development. But in April 1917 there existed only the most rudimentary beginnings of an integrated system for aeronautical research and development.

There was no room for false pride regarding aviation in the United States when the Bolling Mission sailed for Europe.[1] In early 1917 there were only about a half-dozen aeronautical engineers in the Aviation Section of the Signal Corps. Several of these officers were actually studying aeronautics at the Massachusetts Institute of Technology when war broke out.[2] There were at the time only 10 or 12 manufacturers of aircraft in the entire United States. Of these, only about half had ever constructed as many as 10 airplanes. None had ever designed a successful type for combat.

1. Lloyd George's charge that pride kept the United States from profiting by the experience of the Allies in design has been demonstrated as erroneous in Chapter 4. See David Lloyd George, *War Memoirs* (6 vols.; Boston, Little, Brown, 1933–37) *5,* 451; and E. S. Gorrell, "What, No Airplanes?" *Journal of Air Law and Commerce* (Jan. 1941).

2. B. Crowell, *America's Munitions; 1917–1918* (Washington, GPO, 1919), p. 240.

The Curtiss Aeroplane and Motor Corporation of Buffalo, N.Y., having constructed a large number of training aircraft for the War Department and for the British, was the only manufacturer with extensive experience. Moreover, Curtiss enjoyed the advantage of the services rendered by a number of British engineers serving on the staff of the corporation.[3]

There could be no question of the unlikelihood of creating an effective aerial force with these limited resources. Even if the War Department had been able to provide the manufacturers with definite objectives in terms of the performance desired (which the department was quite unprepared to do in April 1917), it is doubtful that a bare half-dozen designers could have developed the types of aircraft required on the front in Europe. The need for aircraft immediately and in large numbers inevitably brought about the decision to turn to Europe for models of tactical aircraft to be put into production at once. On the other hand, the decision to establish an organization for experimental engineering within the military service came only after leaders in aviation circles became aware of the problem of continual change in design. When the Bolling Mission went to Europe for models of aircraft to put into production, the assumption apparently had been that those chosen would be sent home, copied, and turned out in great quantities. It was not until the midsummer and fall of 1917 that the full implications of the never-ending change in design imposed by aerial combat with an aggressive enemy began to affect planning for aviation in the United States.[4]

Not until after the return of aircraft designer Lt. Col. V. E. Clark from his service in Europe with the Bolling Mission did officials in the United States set up an organization to handle engineering problems at the operating level. In October 1917 the Chief Signal Officer authorized the establishment of a Signal Corps "experimental factory" at McCook Field, Dayton, Ohio. Lieutenant Colonel Clark took command of the new organization which was officially designated the Airplane Experimental Department. This position probably curtailed Clark's effectiveness as an aircraft designer by loading him with administrative routine, but there were other reasons why the engineering organization failed to turn out designs for combat aircraft.[5]

3. *Ibid.*, pp. 251–4.
4. See, for example, the gradual change in attitudes of members of the Bolling Mission, Chapter 4, above.
5. Photostat copy, Signal Corps, Equip. Div., Office Memo, No. 30, 18 Oct. 1917, in files of Wright Field Hist. Office. As early as 24 May 1917 an Aircraft Engineer-

The Airplane Experimental Department had scarcely been established when it was confronted with a rival organization, the Production Engineering Department. The Production Engineering Department was originally formed as an agency to prepare and expedite production of aircraft from European sources as well as those conceived by the experimental department, but in actual practice the production department became an experimental engineering agency itself. Many of the designs sent from Europe had to be altered significantly to fit available engines, and in effecting the necessary modifications, the staff of production engineers became experimental engineers perforce. During August 1918 experimental and production engineering combined under a single head.[6]

The urgency of the need for aircraft in France undoubtedly tended to divert the limited number of designers available at Dayton from original work in design to expediting production. Moreover, the staff at McCook Field found that designing aircraft was difficult, if not impossible, when separated from the actual process of manufacture.[7] Whether it was recognition of this difficulty or the pressure of the contemporary need which induced the engineering staff to abandon the attempt cannot be determined. In any event, after the first few months of 1918, the initiative for creative designing of aircraft was left for the most part to the engineering staffs of the several manufacturers.[8]

In view of the limited resources of engineering skill available to the manufacturers of aircraft in the United States, the results achieved in terms of actual performance when war came were remarkable, and at the time of the Armistice a number of indigenous

ing Division had been formed but only as a staff office in the Office of the Chief Signal Officer in Washington. For evidence of the diversion of design talent to command functions, see *Aircraft Production*, Hearings before subcommittee of Senate Military Affairs Committee, 65 Cong. 2 Sess., 15 July 1918, *2*, 779.

6. Some of the problems encountered in converting foreign designs for use with Liberty engines are discussed in the following chapter. See also, for details of organizational changes, Capt. H. H. Blee, "History of . . . Airplane Engineering Division . . . ," 15 Aug. 1919, Wright Field Hist. Office files.

7. For an illustration of this problem, see AAF Historical Study No. 54, "The Development of Aircraft Gun Turrets in the AAF; 1917–1944," p. 280, May 1947, AAF Archive.

8. See "Rotary-Wing Aircraft in the Army Air Forces: A Study in Research and Development Policies," chap. ii, 1946, Wright Field Hist. Office, for evidence of attempts by the government's engineers to design military aircraft. See also J. A. Beck, "Investigations," Sept. 1919, NA, BAP Hist. Box 8, 333.5 Investigations, General; and AAF Historical Study No. 50, "Materiel Research and Development in the Army Air Arm; 1914–1945," chap. i, 1946, AAF Archive.

designs were just at the point of entering full-scale production. The Packard Motor Car Company used a French aeronautical engineer trained at St. Cyr to produce a two-place pursuit, the Lepere, with performance somewhat superior to that of the aircraft in action along the front. The Thomas-Morse one-place pursuit, the Martin bomber, and the Loening two-place pursuit were distinctly domestic designs which had achieved superior performance records when the war ended.[9] But the superlative achievements of purely experimental types did not win battles. Results, no matter how remarkable, were of little value unless they represented aircraft actually on the fighting front. No aircraft entirely designed in the United States reached Europe in time to be of value in World War I.

The achievements of the United States in creative design and experimental engineering, as contrasted to the result in production engineering, were important only insofar as they marked the growth of a new industry and developed a body of experience to guide the War Department in the postwar era. In terms of providing superior weapons to give an advantage for victory, creative design in the United States played little or no part. The effort of the War Department and the aircraft industry together was primarily in the direction of applied rather than creative design. The greater part of the wartime program for aviation was devoted to manufacturing from ready-made designs and to reducing foreign designs to production. Hence there was need for a domestic agency to explore the many horizons of the new science of aviation. Such an agency already existed in the National Advisory Committee for Aeronautics (NACA), but the NACA was not a part of the War Department. The degree to which the War Department would profit from the services of the NACA in perfecting the air weapon was thus largely dependent upon the ability of the department to establish an effective system of liaison with the aeronautical agency.

The NACA has already been mentioned in an earlier chapter as the product of a rider on a naval appropriations bill in 1915. The founding legislation of the NACA grew out of the efforts of a group of enthusiasts carrying on the Langley tradition at the

9. There is no really comprehensive published discussion of aircraft designed in the United States during World War I. Crowell, *America's Munitions,* pp. 253–64, is probably the best available in print. However, for cursory published résumés, see Maj. Gen. G. O. Squier, "Aeronautics in the United States . . . ," address before American Institute of Electrical Engineers (New York, 1919); and Col. G. W. Mixter and Lt. H. H. Emmons, *United States Army Aircraft Production Facts* (Washington, GPO, 1919).

Smithsonian Institution in Washington.[10] The wording of the organic act of the committee constituted a broad directive, which authorized it to supervise and direct the scientific study of the problems of flight. Moreover, when laboratories and equipment became available, the committee was further authorized to "direct and conduct research." The only specific limitation imposed upon the new organization was fiscal, since Congress provided for no more than $5,000 per year for five years.[11]

In creating the NACA, Congress followed a well-established tradition for independent agencies, granting broad, ill-defined powers in the enacting legislation and leaving to time, the initiative of the agency, and the interpretations of the courts the task of defining the agency's functions in detail. It was the aggressive initiative and enthusiasm of the original committeemen on the NACA which determined the real scope of their activities. This delineation of functions began soon after the agency was first created.

During 1915 the committee conducted a survey of facilities for aeronautical research in the United States. The results were disheartening. Few universities displayed more than a "curiosity" in regard to aeronautics; only two offered formal course work in aerodynamics. Thus, in its first months the committee realized that the initial contribution of the NACA would be to define the problem of aviation by isolating the areas of the unknown and grouping them into related fields of study such as power plants, airfoils, instruments, structures, and the like. Having done this, the NACA could then provide a clearinghouse for information and a central source of publication for aeronautical findings.[12] Had the committee done nothing more than provide this medium of expression, it would probably have justified itself in the eyes of Congress, but the men who served on the committee were not content to play a passive role.

The infant aircraft industry of the United States would, the committee felt, follow rather than lead. Without capital and engineering talent, manufacturers had already shown a tendency to follow the demand of the public or the requirements of governmental specifications. It has been suggested (Chapter 2) that before 1917 the leadership of the War Department in aeronautics, while probably somewhat better than it is popularly supposed to

10. *First Annual Report of the National Advisory Committee for Aeronautics* (Washington, GPO, 1916), p. 9. See also AAF Historical Study No. 50, chap. i.

11. Public No. 271, 63 Cong., 3 March 1915. See also Chief Clerk, NACA, to Chief Clerk, Office of Chief Signal Officer, 27 Oct. 1917, AFCF, 334.8 NACA.

12. NACA *Annual Report,* 1915, pp. 12–15.

have been, was scarcely dynamic. Officials of the NACA believed, with ample justification, that aircraft manufacturers were not interested in models involving "radical or sudden changes" from the prevailing standards. Clearly, the committee reasoned, it was the responsibility of a disinterested agency such as the NACA to lead the way in pushing back the horizons of aeronautical science and to set the pace in developing better aircraft.[13]

Beneficial as the initiative of the NACA may have been in stimulating aeronautical research, aggressiveness was not without its more questionable aspects. When the committee turned from research to participate in planning production, it contributed heavily, albeit unintentionally, to the organizational split between operations and supply which did so much to hamper the program of aircraft in World War I.[14] To condemn the committee for the problems resulting from its engaging in activities beyond those contemplated by Congress is to overlook the factor of personality in the new agency. The enthusiasm which led the NACA to undertake planning production in the absence of concrete activity in that direction by the Signal Corps was the same enthusiasm which led the committee to effective service in the field of research.

Officials of the new agency were certainly aggressive. During 1916, the second year of its existence, the NACA secured an appropriation of $85,000 and asked Congress to provide $107,000 for the next fiscal year. Although Congress was doubtless influenced by the progress of aviation in Europe in granting these increases, they still reflect something of the energy with which the committee tackled the problem of aeronautical research. Another more specific index of aggressiveness in the NACA appears in the work of the committee in sending representatives to Europe to obtain scientific information directly from the combatant Powers as early as March 1917. It will be recalled that the War Department, with all its established channels for liaison through military attachés and diplomatic representatives, had been unable to get observers to the front to gather information on developments in aircraft until just about this same time.[15]

13. *Ibid.*, p. 13.
14. NACA *Annual Report*, 1917, p. 17. The NACA put itself on record as basing its proposal for an Aircraft Production Board under the Council of National Defense upon a study of European experience and practice. Unhappily, the NACA doesn't appear to have appreciated the lack of full ministerial responsibility in the Council of National Defense as contrasted with the British Ministry; for example, the proposal of the NACA adopted the duties but not the authority, the form without the substance, of its European models. See above, Chapter 4.
15. NACA *Annual Report*, 1916, pp. 17–19; and Extract from draft of NACA

When the United States finally entered the European war, the NACA expanded its activities to meet the special problems raised by the use of aircraft in warfare. Thus, by the end of 1917, regardless of whatever might have been the intention of Congress, the NACA was performing four relatively distinct functional roles: encouragement to civil aviation; dissemination of aeronautical intelligence or technical information; applied research with special military ends in view; and fundamental research. Of these only the last fitted in with the role probably intended for the agency by Congress.[16]

During the first two years of its existence the NACA operated from a few crowded rooms in a Washington office building. During 1916 a site board tried to find a suitable location for a research center. When at last the War Department was persuaded to provide a plot at Langley Field, Va., work began on a laboratory building and wind tunnels for which nearly $90,000 had been appropriated.[17] At the end of the war, the wind tunnels were still incomplete. The program of fundamental research by the NACA may have had little or no appreciable influence upon the outcome of the war, but the stimulus of war, especially in terms of funds, made it possible to build up facilities which provided invaluable services to the air arm during the next 20-odd years.[18]

Of equal importance, or perhaps of greater importance as far as this study is concerned, were the administrative relationships which the NACA worked out with the War and Navy departments during the years leading up to the Armistice. Both the War and Navy departments were granted an element of control in the NACA by the organic legislation of 1915 which gave each service two members on the Executive Committee. With representatives on this policy-making committee, it was assumed that the two departments would be able to influence the direction of the program of research conducted by the NACA. This they did in later years, but during World War I there was, as it has been shown, no laboratory and very little research, so control of policy was academic.

Executive Committee report, F.Y. 1916–17, AFCF, 334.8 NACA. For a comparison of NACA and War Department in securing information from Europe, see above, Chapter 3.

16. A detailed list of the subcommittees, their personnel, and functions is given in NACA *Annual Report,* 1917. See also S. W. Stratton, NACA, to Capt. T. D. Milling, Aviation Sect., Signal Corps, 15 May 1917, AFCF, 334.8 NACA.

17. NACA *Annual Report,* 1917, p. 16.

18. AAF Historical Study No. 50, pp. 14–15. See also NACA *Annual Report,* 1918, p. 24.

Soon after the United States entered the European conflict a flood of "inventions" from all over the nation poured in from persons who wished to contribute to the war effort. Up until the time Gen. P. C. March became Chief of Staff, inventions which were offered to the War Department were referred to the old Board of Ordnance and Fortification, an agency whose ineptitude for the wartime task is probably best indicated by the fumbling form-letter treatment the board had given the Wright brothers in the leisurely days of peace. General March abolished the board as a "sheer waste of time." In its stead during July 1918 he established on the General Staff an Inventions Section manned with scientists and specialists drawn from universities for the duration.[19]

Aircraft inventions, on the other hand, presented a quite different problem. Long before the Inventions Section of the General Staff appeared, the NACA had begun to function as a clearing-house for ideas, screening literally thousands of proposals before sending on a small percentage of useful inventions to naval and military engineers.[20] It was in performing this useful function that the NACA hammered out its working relationship with the War Department. The two formal representatives of the War Department on the Executive Committee were officials from Washington who attended meetings held at infrequent intervals, but the day-to-day relationship of the department and the NACA was actually carried on by officials of the NACA and members of the engineering staff in echelons of the service far below the officials responsible for effective policy-making.

The practical result of the difference between official representation and operating relationships was to shape the character of the influence which the NACA had upon the development of military aircraft in the postwar years. By submitting ideas directly to the engineering staffs, the NACA placed itself in the hands of those staffs. During the war, for example, one innovation in the construction of aircraft which came to the attention of the NACA was referred to the engineers at McCook, where the idea was either rejected or ignored. The same idea was given to the navy and proved to be of outstanding value.[21] The structure in question may have been relatively unimportant in terms of the broad picture of devel-

19. P. C. March, *The Nation at War* (Garden City, Doubleday Doran, 1932), p. 47.

20. NACA *Annual Report*, 1918, pp. 29–30.

21. Report on inventions handled by NACA, typescript account presumably prepared by officials of the NACA, 13 April 1918, AFCF, 334.8 NACA.

opment in the field of aircraft, but the administrative relationships the incident betokened were important.

Submitting ideas through engineering channels rather than through policy planners in the topmost echelons made it possible for the engineering staff of the air arm to kill off a proposal for a potentially good weapon before it reached the attention of those who shaped strategy. It is very probable that few if any revolutionary weapons were "lost" by the services in this manner during World War I, but the subservience of the committee to engineering rather than strategy was apparently firmly established by the wartime precedent.[22]

It has already been suggested that the creation of the NACA offered the War Department an important means of improving the air arm without assuming budgetary responsibility if it could evolve an effective system of liaison for exploiting to the utmost the capabilities of the new agency. If the system of liaison was ineffective, the fault was not entirely military. The committee lacked singleness of direction. To be sure, by 1918, after the aircraft program had failed spectacularly and had been subjected to numerous investigations, the committee was only too willing to report that its activities no longer included planning programs of production.[23] But even when the agency had restricted itself to the role of a research organization, responsibility was not clearly defined until after the Armistice when the committee made a determined effort to clarify its role.[24] All temporary or expedient wartime subcommittees were discharged in 1919, and the agency's function was redefined along twofold lines as a clearinghouse for all aeronautical information and a research agency to service the various branches of the government.[25]

The NACA spent most of the war years in finding itself. The problem was by no means unique. Other scientific agencies experienced comparable troubles in trying to work out effective systems

22. The postwar practice was to appoint the Chief of the Air Service and the Chief of the Engineering Division as representatives of the army on the NACA. Although this arrangement might seem to have balanced the strategic factor *vs.* the engineering factor in determining the course of NACA research, correspondence for the period 1920-42 shows the predominance of the engineering influence. See WFCF, 334.8 *passim.* See also Copy, memo, J. C. Hunsaker, NACA, to Gen. H. H. Arnold, 2 May 1942, AFCF, 334.8 NACA.

23. NACA *Annual Report,* 1918, p. 25.

24. For an example of some internal rumblings in NACA, see L. M. Griffith, Senior Staff Engineer, to Exec. Com., 4 Sept. 1918, AFCF, 334.8 NACA.

25. NACA *Annual Report,* 1919, p. 11.

of liaison with the War Department. To consider some of these
other agencies it will be necessary to revert briefly to the early
months of the war.

In April 1917 the National Academy of Sciences offered its
services to the government of the United States in organizing for
war purposes the scientific resources of the country: educational,
industrial, and institutional. The offer was a logical one inasmuch
as the National Academy of Sciences was originally chartered
during the Civil War to abet the Federal cause by mobilizing
scientific intelligence. The National Academy's charter was a broad
mandate to investigate, examine, experiment, and report upon any
subject of science when called upon to do so by any department of
the government. Within the framework of this charter, the National Academy of Sciences had shown commendable initiative in
preparing for war before war was declared.[26]

As early as February 1917 the academy secured official approval
for creating the National Research Council (NRC), an organization primarily designed to ferret out and make available the
nation's scientific personnel. Something of the council's initiative is
shown in the energy with which it organized an expedition of scientific observers to the French front a full month before the United
States entered the war.[27] Unfortunately, initiative and even aggressive interest were insufficient substitutes for effective organization. From the outbreak of war until early in 1918 the NRC existed only as a relatively loose assembly of scientists. A single representative in Washington served in the capacity of liaison officer
between the federal government and individual scientists working
in their home localities, almost entirely without benefit of government funds. Limited as the productivity of this rather informal
arrangement necessarily was, the NRC nevertheless laid some important groundwork in organizing science for war. During the
summer of 1917, at the council's instigation, the Chief Signal Officer established a Science and Research Division within the Signal
Corps which utilized scientists of the NRC to carry out a number
of research projects. The distinguished physicist R. A. Millikan
accepted a commission as lieutenant colonel and took command of
the division in 1918.[28]

26. W. F. Willoughby, *Government Organization in War Time and After* (New
York, D. Appleton, 1919), p. 22.

27. NACA *Annual Report,* 1917, p. 19.

28. Capt. C. M. Sparrow, "A Brief History of the Organization and Activities of
the Science and Research Division of the Bureau of Aircraft Production" (undated,
*ca.* 1919), NA, A.S. Finance Advisory Board, Case No. 310, Box 15. See also Lt. R. C.

The exact role of the Science and Research Division in the administrative framework of the Signal Corps was not clearly established. The directive which authorized the new division vaguely assigned it responsibility for all research and development work ordered by the Chief Signal Officer. This left the entire burden of initiative upon the chief and none with the scientists themselves. Under this arrangement the scientists of the division were considered to be a pool of talent from which the best qualified individuals were drawn when problems appeared. At best this was an after-the-fact way of exploiting the potentialities of science, waiting for difficulties to develop rather than seeking to use scientists to develop difficulties—for the enemy.

Within the limitations imposed by a scarcity of funds and the absence of effective liaison in Washington, the scientists of the NRC proved sufficiently useful during the summer and fall of 1917 to induce officials in the Signal Corps to take increased interest in the possibilities of exploiting the scientific resources of the nation. Early in 1918 the Chief Signal Officer issued a new directive strengthening the position of the Science and Research Division by giving it a more aggressive role. This directive made the division responsible for the investigation of all inventions and scientific developments of interest. In place of passive reliance upon orders from the Chief Signal Officer assigning individual projects for solution, the scientists were now officially encouraged and even required to seek out technological advances of possible use for military purposes. Of equal or greater importance, the new directive specifically required the Science and Research Division to bring findings of interest to the attention of other divisions of the Signal Corps which might make use of them.[29]

The scientists assigned to work with the Signal Corps were, of course, only one group of the many enlisted in the over-all program of the NRC which served all branches of the War Department, the Navy Department, and other governmental agencies. Nonetheless, the success of the Science and Research Division in the Signal Corps appears to have been typical of other units in the NRC with military affiliations, because President Wilson signed an executive order in May 1918 requesting the National Academy of Sciences

Hillsdale, "Advisory and Cooperative Agencies," pp. 57-70, July 1919, NA, BAP Hist. Box 9, 334.7 NRC History.

29. Sparrow, "A Brief History"; and Office, Chief Signal Officer, Office Memo No. 35, 16 Feb. 1918, NA, DMA Box 3, 321.9. See also Memo, Lt. Col. L. S. Horner to Maj. R. A. Millikan, Chief, Science and Research Div., 28 Feb. 1918, NA, BAP Exec. Box 6, 026.4 Scientific Research.

to give the NRC permanent status. The president laid down a series of specific functions to guide the council in its future operations. The broad purpose of the NRC was to stimulate science and research for national defense. Within the meaning of this general objective the council was enjoined to "survey the larger possibilities of science" and to "formulate comprehensive projects," bringing the attention of scientists and technical investigators to the important requirements of the armed forces.

The president's directive for the NRC was a great deal more specific than that guiding the NACA. In a sense, it reflected the experience of the Science and Research Division in the Signal Corps, for it required a positive or active role rather than a passive one. Unfortunately, while the directive was specific in telling exactly what was wanted, it neglected the important consideration of how the desired end should be accomplished. The NRC was "to serve as a means for bringing . . . investigators, into active cooperation with the scientific and technical services of the War and Navy Departments . . ." [30] Implicit in this assignment was the assumption that the military would come halfway, organizing an administrative system capable of working in conjunction with the scientific talent made available by the NRC. Just as in the case of the NACA, the extent to which the War Department profited from the services made available was, in a large measure, dependent upon the effectiveness of the machinery for liaison established to bring the two organizations together.

The Chief Signal Officer's directive of February 1918, mentioned above, gave the Science and Research Division an active rather than a passive role. Three months afterward the Signal Corps lost all control over aviation with the formation of the Division of Military Aeronautics (DMA) and the Bureau of Aircraft Production (BAP). The Science and Research Division moved under the control of the BAP. But a complete break with the Signal Corps proved difficult. Many of the scientists of the division were engaged in work on signal equipment rather than aircraft, so the division continued to serve the Signal Corps as well as the newly created organization for producing aircraft.[31]

Shortly after the reorganization in May which divested the Signal Corps of aeronautical interests, the Secretary of War redefined

---

30. The full text of the Executive Order of 11 May 1918 is given in Willoughby, *Government Organization*, pp. 23–4.

31. Undated, unsigned typescript, "Science and Research Division," probably prepared sometime in 1919, NA, DMA Box 3, 321.9.

the duties of the Science and Research Division. In the light of the Chief Signal Officer's directive of February, the secretary's concept of the functions of the division appears to be a retrogression. He believed the division should be "utilized for such scientific investigation and research as may be necessary." Such a definition outlined a decidedly narrower and more passive role than that contemplated by the Chief Signal Officer.[32]

The Science and Research Division had been operating under the BAP for less than three months when it became evident that the new arrangement was not so effective as might be desired. The Director of Technical Information (BAP) recognized the problem for what it was. Research work on aeronautical problems, he felt, had been unsatisfactory, in part at least, as a direct result of inadequate organization and administration. Not only were the internal operations of the Science and Research Division unstable, but the relationship and partition of responsibilities among the divisions, the NACA, and the Bureau of Standards, as well as a number of other governmental agencies remained unclear.[33]

The Director of Technical Information was in an excellent position to understand the limitations imposed upon the Science and Research Division while its organizational stature remained uncertain. Sitting in the midst of a complex where competitive and duplicating channels of information were the order of the day, the director made a determined effort to unify the research activities of the Science and Research Division, the NACA, and other agencies under a single command. The need for unification and control was probably not so much the result of an abuse of power by those who planned and created the several agencies for research as it was the result of a combination of carelessness in drafting directives and innate aggressiveness on the part of operating personnel. If it was to avoid competition and collision with other research agencies, the organization, with its greatly expanded scientific and research functions, now required better administrative control than it had enjoyed when still on a job-shop or project basis under the Chief Signal Officer. Enlarged functions tended to penetrate areas of activity already covered by parallel agencies, making a greater refinement of directives and a tightening of administrative control increasingly necessary. It was this situation which the Director of Technical Information, with his sorry experience concerning the

32. Sparrow, "A Brief History."
33. W. C. Sabine, Director of Tech. Info., BAP, to M. W. Kellogg, Asst. Director, Aircraft Production, 20 July 1918, NA, BAP Exec. Box 6, 026.4 Scientific Research.

problem of technical data, recognized so clearly. Seeing the problem and doing something about it, though, were two different matters.

The several research organizations interested in aeronautics represented as many different and separate governmental agencies, parallel to the air arm proper rather than subordinate to it. Effective reorganization would involve more than administrative reform within the War Department; possibly even legislative action would be necessary. Unfortunately neither Congress nor the War Department attempted any comprehensive solution of the problem of administering research. The Director of Technical Information felt that "a very large opportunity to accomplish results by suitably organized and directed research" had been thrown away.[34]

Not only was there no attempt to unify aeronautical research activities at the national level in the summer of 1918, but even within the BAP itself the Science and Research Division appeared to be relegated to an inferior status. An administrator for industrial research recruited as a possible director for a unified organization believed that research was subordinated to production in the BAP. After studying the problem in some detail, he reported that the production officials were "probably not very keen on scientific research" and were "preoccupied by immediate difficulties." [35]

When the war ended, production was still considered more urgent than scientific research. While there were literally thousands of employees administering production of aircraft at the time of the Armistice, that portion of the Science and Research Division concerned with aviation comprised a total of 22 officers, 121 enlisted men, and 16 civilian scientists.[36] Numbers alone, of course, were no certain indication of significance, but, the fact that a large proportion of the scientists, civilians and officers alike, had been drawn from outside the War Department for the emergency only was of utmost importance. When the war ended they went home to the universities and research institutes from which they had come. The Science and Research Division almost ceased to exist for want of personnel.

There were other reasons too why the Science and Research Division did not survive the immediate shock of the end of the war. It

---

34. W. C. Sabine to W. R. Whitney (of General Electric, Schenectady, N.Y.), 22 Aug. 1918, and Memo, Asst. Director, Aircraft Production, to Bureau of Standards, 12 July 1918, NA, BAP Exec. Box 6, 026.4 Scientific Research.

35. W. R. Whitney to W. C. Sabine, 27 Aug. 1918, NA, BAP Exec. Box 6, 026.4 Scientific Research.

36. Sparrow, "A Brief History."

had not become deeply entrenched in the organizational hierarchy of the air arm. To begin with it had been an extraneous accretion to the Signal Office, and after the reorganization of May 1918 it straddled uncomfortably between the Signal Corps and the BAP, which was more interested in production than in research. When the emphasis on production fell away after the Armistice, the BAP was dissolved in the formation of the Air Service, and the Science and Research Division was nominally merged with the engineering organizations at Dayton, Ohio. Without the leadership of its scientific personnel the division changed in character as it became subservient to an engineering agency with different objectives.[37] Together these factors added up to one result: Little interest and less experience in organizing for science and research were passed on to the postwar era by the World War I agencies for military research. Since these agencies made comparatively little direct contribution to victory, the United States necessarily depended heavily upon borrowed weapons and foreign designs for aircraft and accessories. This reliance served to emphasize the importance of the organizations for "borrowing," to wit, the organizations for information and decision-making discussed in earlier chapters. Just how well they were able to carry out their functions remains to be seen.

37. See "Rotary-Wing Aircraft" (cited above, p. 105 n. 8) for a discussion of the role of fundamental research in an applied research or engineering organization.

## Chapter VII. The Development of Air Weapons: Fighters and Observation Aircraft

THE UNITED STATES declared war on Germany in April 1917. A full year later aircraft production amounted to little more than a trickle. No bombers and no fighters had been produced. The first nine completed observation aircraft, modified versions of the British DeHaviland DH-4, appeared in February 1918. During the following month only four more came off the assembly lines, and in April, a full year after the war began, the total production was 15 aircraft.[1] Of the several foreign types selected by the Bolling Mission in the summer of 1917, only one had even approached the stage of mass production. Although there were 10,000 DH-4 observation aircraft on contract during the war, 15 completed aircraft some 12 months after the outbreak could scarcely be regarded as an impressive achievement.[2] The promised aerial might with which the United States hoped to darken the skies of Europe had not materialized.

When the public at large learned of the insignificant number of aircraft manufactured, a wave of indignation and criticism broke over those who were responsible for the nation's aviation program.[3] Just as might have been expected, a rash of investigations, official and quasi-official, broke out in answer to the public clamor. The story of these investigations and the factors, political as well as patriotic, which inspired them is not the concern of this study. Some of the investigators' findings, nonetheless, have a direct bearing upon the development of superior aerial weapons. The successive investigations uncovered a great number of flaws in the program

1. Col. G. W. Mixter and Lt. H. H. Emmons, *United States Army Aircraft Production Facts* (Washington, GPO, 1919), p. 48.

2. *Report on Aircraft Surveys* . . . , House Document No. 621, 66 Cong. 2 Sess., 19 Jan. 1920, p. 4.

3. See, for example, New York *Times,* 20 March 1918 ff., for criticism of the aviation program.

for manufacturing aircraft, but the consensus held that inadequate organization was the root of the trouble.[4]

A subcommittee of the Senate Military Affairs Committee, one of the last of the investigating groups, published its report in August 1918. The committee's findings were in many ways a summation of the investigations conducted earlier, and as such they were generally representative. The Senate committee was quick to point out that an "unsystematic and ineffective" organization in which the several branches suffered because of ill-defined, conflicting, and overlapping functions had hampered production from the beginning of the war. The administrative deficiencies of the organization for the production of aircraft were reflected in a number of ways. The ineffective systems of liaison to report the requirements of the forces in combat to the drawing boards of industry were singled out for particular criticism. The committee also emphasized the need for a single, unified source of command to give authoritative decisions in the selection and development of aerial weapons.[5]

The findings of the Senate committee, like those of all the other investigating groups, uncovered much that was faulty, but nowhere did they resolve the various difficulties into a comprehensive explanation of the problems encountered while developing aircraft in wartime. Nowhere did the committee attempt to relate the inadequacies of the system of liaison or the faulty organization for making decisions directly with the meager results accomplished. The committee did go as far as to place the major blame for the failure of the program upon the policy of adapting all tactical aircraft to the Liberty engine rather than manufacturing exact copies of European models. Nevertheless, the factors behind this all-important policy and their relationship to the inadequate organizations for information and decision were left unexplored, if indeed they were even recognized.

In singling out for censure the decision to standardize the Liberty engine for all tactical aircraft, however, the Senate committee might well have contributed substantially to an understanding of the whole problem. The case of the Liberty engine is essentially the pattern of aeronautical design in brief. To analyze the decision

4. For comparison of the investigating committee texts, see reports of H. Snowden Marshall, 12 April 1918; the Senate Committee, 22 Aug. 1918; and Charles Evans Hughes, 25 Oct. 1918, NA, BAP Hist. Box 8, 333.5 Investigations, General.

5. *Aircraft Production in the United States*, Senate Report No. 555, 66 Cong. 2 Sess., 22 Aug. 1918, pp. 3–4.

to standardize the Liberty is to analyze the whole wartime program
for developing aircraft.

When the United States declared war there were no satisfactory engines in the United States suitable for use against the enemy.
Foreign aviation missions in Washington during the spring of
1917 agreed that a 225 h.p. engine should be developed for use in
the following year. Three months later it was evident that a 330 h.p.
engine would be required if aircraft of the AEF were to compete
successfully with the enemy.[6] The Liberty engine was initially designed to meet the 225 h.p. requirement. Before the war ended the
original 8-cylinder engine had been expanded into a 12-cylinder
model. The power rating of the Liberty 12 jumped in several incre-
ments from 330 h.p. to a final model at the end of the war rated at
440 h.p.[7] Each increase in power involved redesign, prolonged testing, and delays in production. At the same time, each additional
increase in horsepower rating widened the margin of superiority
in combat.

Almost from the very beginning of the war it was clear to officials in the United States that the pace of increases in horsepower
would be the pace of development in aircraft. The decision to
utilize the Liberty engine, an admittedly experimental type, was
a decision made with full appreciation of the necessity for designing well ahead of the course of events in Europe. To build exact
copies of the foreign types available when the United States declared war would have meant providing obsolete aircraft for the
AEF. The time involved in transmitting drawings and physical
samples from Europe and the time spent in getting production
under way in the United States would inevitably amount to at least
several months. That the 225 h.p. engines recommended for production in April and May of 1917 were considered inadequate a
mere three months later is in itself evidence of the wisdom of the
decision to use Liberty engines rather than foreign models. Furthermore, the decision of the authorities to use the standard Liberty
engine for all types of tactical aircraft manufactured in the United
States, or whenever possible, was a decision to foster output in
quantity. Standardization is the essence of mass production. To
standardize is to simplify. Not only does standardization simplify

6. Mixter and Emmons, *Aircraft Production Facts*, p. 16 n. 1. At the time the
United States entered the war, Allied engines then in use seldom developed more
than 150 h.p. The British were even then striving to produce 200 h.p. engines in
quantity. H. A. Jones, *The War in the Air* (5 vols., *2–6*; Oxford, Clarendon Press,
1928–37), Vol. *6*, chap. ii.

7. Mixter and Emmons, *Aircraft Production Facts*, pp. 17, 21.

tooling, training of labor, and manufacturing operations, but maintenance in the field and distribution of spare parts are vastly facilitated as well.

Thus far, the officials were consistent. Their decisions showed a recognition of the importance of superiority in design and at the same time the equally important factor of production in quantity. The problem of producing superior airplanes in wartime was, unfortunately, not quite so readily solved. As all designers knew, airframes are planned around engines. The power plant is the heart of the aircraft. To standardize with one engine was to force all designs to conform to the limitations and characteristics of that one engine regardless of the functions to be performed by the airplane. When a standardized engine was imposed upon designers of aircraft, long-range night-bombers were limited to the same power plant used by low-flying observation aircraft. To standardize was to stultify creative design and the development of aircraft as a whole.[8] The investigating committee of the Senate recognized the facts of this situation but failed to take the next logical step and resolve these facts into an explanation of the underlying difficulty during the war.[9]

The problem of the Liberty engine was typical of the larger problem of aircraft production as a whole. Standardization and superior weapons involve a conflict of policies representing mutually exclusive ends. The objectives of more weapons and of better weapons tend to pull in opposite directions. The extent to which officials in charge of aviation recognized the contest of objectives would determine, to a large extent, the results achieved in manufacturing aircraft for the AEF.

By the end of 1917 officers in the War Department at large appeared to recognize the presence of the forces of both standardization and development. "Aviation science," the Chief of Staff reported, "is advancing by leaps and bounds." The aircraft of "today," he said, is obsolete "tomorrow." At the same time the chief praised the Bureau of Standards for perfecting a scheme of standardization which avoided "the absurd condition previously

8. The stultifying influence of the standardized Liberty engine was somewhat mitigated by the circumstance that other engines actually were developed and manufactured in the United States in spite of the policy of standardization. During the 1920's Liberty engines in surplus stock retarded aircraft development so seriously that Congress finally arbitrarily forbade their use. See AAF Historical Study No. 44, "Evolution of the Liaison-type Airplane, 1917–1944," chap. i, May 1946, AAF Archive.

9. *Aircraft Production in the United States*, Senate Report No. 555, p. 3.

existing" wherein thousands of different parts for every type of aircraft clogged the industries of the nation.[10] In his final report at the end of the war the Chief Signal Officer, if not the entire War Department, was ready to attribute the disappointments in production to the failure to standardize. One of the "serious mistakes" of the war, he believed, was the "multiplicity of types" developed. On the other hand, he lauded the whole conception of a single standard engine as an outstanding achievement. The celebrations of October 1918 in Detroit, when the ten-thousandth Liberty engine rolled off the production line, appeared to give the Signal Corps a sense of great accomplishment.[11] It may be argued that the report was pleading a cause inasmuch as the Chief Signal Officer had been involved in the decision to standardize the Liberty. But if the Chief Signal Officer's sense of triumph was special pleading, his case was not without support.

Throughout the first six months of 1918 the demands from the front were for aircraft in quantity; they did not stress quality. In July 1918 the Chief of the Air Service, AEF, laid down a policy which expressed in a single sentence the point of view of the fighting front more succinctly than a dozen cabled discussions: "Improvements are good but production is better." [12] The sense of urgency raised by repeated demands from the front tended to place an abnormal emphasis on production and hence upon standardization at the expense of improvement in design. Had the War Department concentrated upon production in quantity, manufacturing exact copies of aircraft and engines from designs on hand in April 1917, there is little doubt that the numbers of aircraft desired on the front would have been available. But if that course had been taken there is equally little doubt that the protests from the front would have been loud and long that obsolete aircraft were worse than none at all, were, in fact, suicidal. Resolution of the contest of weapons of superior performance *vs.* weapons standardized for

10. Report of Chief of Staff, *Annual Reports of the War Department,* 1917, *1,* 137.

11. Report of Chief Signal Officer, *Annual Reports of the War Department,* 1918, *1,* 1075–9.

12. A.S., AEF, Memo No. 21, 15 July 1918, quoted in H. A. Toulmin, *Air Service, American Expeditionary Force, 1918* (New York, D. Van Nostrand, 1927), p. 144. Toulmin, although he had access to the facts, utterly fails to understand the problem of performance *vs.* production and repeats Lloyd George's charge that pride led the United States to develop the Liberty. Toulmin places the blame upon "personal vanity of designers and production men" and a "provincial but natural pride," p. 143.

mass production was never so simple as the critics on the front seemed to think.[13]

Months before the United States entered the war the British had learned the lesson that standardization brings stultification. The War Office, influenced no doubt by officials at the Royal Aircraft Factory at Farnborough, standardized on the 75-mph BE, or Blériot Experimental aircraft, as a type for combat. As a result of the refusal to accept or at least encourage competitive development by private manufacturers, the 75-mph British airplane became easy prey, "Fokker Fodder," when the Germans appeared along the front with the 100-mph Fokker aircraft. By 1916 the War Office knew the price of standardization and thereafter actively encouraged the development of new designs.[14] The British had learned, as Chief Justice Holmes once said, "To rest upon a certainty is a slumber which, prolonged, brings death."

Standardization may be the birth of production, but it is at the same time the death of development. As one officer in the AEF described the condition, the types of aircraft required at the front "changed more rapidly than women's millinery." The following chart, indicating the number of different types of aircraft developed by the Powers during World War I to fulfill three basic military functions of the air arm, clearly reveals the high rate of flux in design in the contest for superior performance.[15]

13. Typical of the combat aviator's attitude was the following complaint: "There was a notable lack of cooperation . . . as shown by the failure of the Technical Section to adopt improvements suggested by commanders who had learned from experience that such improvements were essential to efficient operations." "Tactical History of American Day Bombardment Aviation," p. 28 (undated, *ca.* 1919), NA, WWI Orgn. Records, A.S. Hist. Records Box 1. While the organization for translating ideas into weapons was certainly faulty, even the best of organizations would have been unable to effect the miracles demanded by some tactical commanders. For an example of the evil influence of obsolete or inferior equipment on morale, see Jones, *The War in the Air, 6, 36.*

14. C. G. Grey, *The History of Combat Airplanes* (Northfield, Vt., Norwich University, 1941), pp. 8–12, and *A History of the Air Ministry* (London, G. Allen and Unwin, 1940), pp. 40–1, 51, 61–2. W. A. Raleigh, *The War in the Air* (Oxford, Clarendon Press, 1922), *1*, 426, 429, and H. A. Jones, *The War in the Air* (5 vols., *2–6;* Oxford, Clarendon Press, 1928–37), *2*, 261, discuss the BE but ignore the full implications of the problem. Jones states (*3, 267*) that three-quarters of the RFC squadrons in France at the beginning of 1916 were equipped with a standardized BE aircraft which was inferior to the Fokker.

15. E. S. Gorrell, "The Council of Errors," address before Virginia Military Institute, 13 Feb. 1941, and "What, No Airplanes?" *Journal of Air Law and Commerce* (Jan. 1941), p. 18. See also "Final Report of the Chief of the Air Service, AEF," *Air Service Information Circular, 2,* No. 180 (15 Feb. 1921), 37–8. The figures shown include different models (representing major modifications of types) as well as different types.

NUMBER OF TYPES DEVELOPED IN EACH OF
THE FOLLOWING CATEGORIES

|  | Observation | Pursuit | Day Bombers | Night Bombers |
|---|---|---|---|---|
| United Kingdom | 20 | 27 | 10 | 10 |
| France | 22 | 31 | 7 | 4 |
| Italy | 11 | 13 | 4 | 7 |
| Germany | 10 | 12 | — | 6 |

In the light of the almost fantastically rapid obsolescence shown above, the standard Liberty engine must be appraised as neither a triumph nor a failure. The Liberty engine epitomizes the problem of aircraft in wartime. Insofar as the Liberty was turned out in quantity—over 15,000 by November 1918—the engine was a success in production.[16] Moreover, since the Liberty 12 was modified to increase its output in horsepower progressively from 330-h.p. to 440-h.p., the engine was a success in design capable of expanding its margin of performance to maintain superiority in combat.[17] The contest between the objective of mass production and the objective of superior or improved performance resulted in a satisfactory working compromise where the Liberty engine was concerned. In the struggle for more weapons *and* for better weapons a balance had to be obtained. The program of the War Department for producing aircraft during World War I was a failure to the extent that it did not attain in production of airframes the same balance between numbers and improvement accomplished with the Liberty engine.

The automotive industry which produced the Liberty engine was well established and fully operating when the war came. The organization which made the decision to use the Liberty engine, or misuse it, had no such foundation. Neither the handful of officers in the Aviation Section of the Signal Corps nor the minute aircraft industry had had at the outbreak of war any significant experience in production, administrative organization, or the development of airplanes for combat. The results achieved were commensurate with the inadequacies of the organization. A résumé of wartime production reveals just how inadequate the organization really was.

The British Bristol fighter was one of the three British aircraft selected by the Bolling Mission for production in the United States. A sample aircraft arrived in New York on 25 August 1917. Ten days later it reached Washington where Lt. Col. V. E. Clark, who had just returned from Europe, made preliminary drawings, re-

16. Mixter and Emmons, *Aircraft Production Facts*, p. 25.
17. *Ibid.*, p. 23.

designing the airframe to use a Liberty engine. The Liberty, rated at 400 h.p. in contrast to the 275 h.p. engine conventionally utilized in the Bristol, was a larger and heavier power-plant and required a considerable amount of modification in the airframe to be adapted successfully.[18] While the redesign of the Bristol fighter was in process, the Equipment Division of the Signal Corps signed a contract for 2,000 Bristols with the Curtiss Airplane and Motor Corporation, at that time the largest builder of aircraft in the United States. During the first week of November 1917, military officials sent both the British sample aircraft and the drawings converting the Bristol for use with the Liberty engine to the Curtiss factory. By the end of the month construction had actually begun on the first item. So energetically was the work pushed that an aircraft was completed for a flight test before the end of January 1918.

The Bristol with Liberty engine proved to have a number of faults which engineers at the Curtiss plant set out to eliminate by redesign. Meanwhile, preparations went ahead for production of the 2,000 items on contract. In their successive attempts to redesign the Bristol to carry the Liberty, engineers of the Signal Corps and Curtiss staff increased the over-all weight of the aircraft. Since the wing area was not changed, the wing-loading, or pounds per square foot of wing area, increased alarmingly. Twenty-six Bristols were manufactured, but those which were given flight tests crashed. During July 1918, the War Department canceled the Bristol contract after spending more than six millions.[19] This loss in money was probably insignificant in comparison with the loss in time. After 15 months of war, no fighter aircraft had reached production in the United States. In converting the Bristol for use with the Liberty engine, aviation officials had gambled in the hope of securing an aircraft of superior performance rather than turning out exact copies of British designs in large numbers. The gamble failed.

The story of the French SPAD, a fighter selected by the Bolling Mission, parallels that of the Bristol. Arriving in the United States in September 1917, the sample SPAD was shipped directly to the Curtiss factory in Buffalo, New York. The manufacturer had no sooner received an order for 3,000 single-place SPAD fight-

18. *Ibid.,* p. 49.

19. O. I. Brockett, "History of the Bristol Fighter," 1919, and documentary data compiled in preparation of this document, NA, BAP Hist. Box 20, 452.1 Bristol Aircraft. See also *Aircraft Production in the United States,* Senate Report No. 555, pp. 2, 5, and *Report on Aircraft Surveys . . . ,* House Document No. 621, p. 4.

ers and begun to prepare for production than a cancellation arrived. The whole project stopped.

A number of factors probably entered into the decision to cancel the order for SPADS. Lieutenant Colonel Clark believed it would be impossible to convert the SPAD to carry a Liberty engine. Also, repeated cables from the AEF questioned the value of single-place fighters.[20] During December 1917 a cable sent from abroad, probably without knowledge of the order canceling the SPADS, recommended that effort in the United States should be concentrated on aircraft "already on our program." A further recommendation that the United States should "leave production of single-place fighters to Europe" served only to compound the confusion and confirm the probable error.[21]

In February 1918, when French sources proved unable to meet the demand, cables from the AEF requested 1,000 SPADS "for earliest possible delivery to France." In April 1918, a full year after the United States entered the war, subsequent dispatches from the AEF directed "immediate preparations for the production of single-seater machines."[22] At the end of April, Curtiss received a contract for 1,000 SE-5 aircraft, the British version of the French SPAD fighter. This meant beginning all over again, starting at the point where the previous contracts to manufacture SPADS had begun eight months earlier. But of course the new model never reached production before the Armistice.[23] The absence of adequate information and machinery for making decisions seemed never more apparent.

When the war ended with neither the SPAD nor the Bristol in production, the United States had no fighter aircraft on the front other than those secured from foreign sources. With the British DH-4, the observation aircraft selected by the Bolling Mission for production in the United States, the achievement was quite different. By the time the Bolling Mission broke up and Lt. Col. V. E. Clark returned home, the DH-4 had already become ob-

20. See, for example, Copy, Pershing cable, 5 Oct. 1917, NA, BAP Hist. Box 25, SPAD file, and Hughes Report, 25 Oct. 1918, *Automotive Industries; The Automobile, 39,* No. 18, (31 Oct. 1918), 745 ff.

21. Copy, Pershing cable No. 375, 14 Dec. 1917, NA, BAP Hist. Box 25, SPAD file.

22. Pershing cables No. 589, 10 Feb. 1918, and No. 916, 19 April 1918, among others, NA, BAP Hist. Box 25, SPAD file.

23. Testimony of Maj. B. D. Foulois, *War Expenditures,* House Hearings, 66 Cong. 1 Sess., Serial 2, Pt. 6, 6 Aug. 1919, pp. 378–9; *Aircraft Production in the United States,* Senate Report No. 555, p. 2; and J. A. Beck, "Investigations," Sept. 1919, NA, BAP Hist. Box 8, 333.5 Investigations, General.

solescent.[24] A subsequent model, the DH-9, was beginning to replace the DH-4 in British production, and Colonel Clark recommended manufacture of the more advanced design.[25] Unfortunately the only sample aircraft available in the United States was a DH-4 and not a DH-9. This sample had actually reached New York before the Bolling Mission finally decided upon the DH-4 as an official selection. But this forehandedness and spirit of cooperation on the part of the British were, as things turned out, to prove a misfortune. During August 1917 the sample DH-4 was shipped to the Dayton-Wright Airplane Company in Ohio where engineers began preliminary studies to convert the aircraft for use with the Liberty engine. When Colonel Bolling learned in France of a plan to manufacture 2,000 DH-9 aircraft for observation, he promptly cabled back that the number should be doubled. By the end of September 1917 it was anticipated that production on 4,000 DH-9 aircraft would begin in about three months. But at that date there was no DH-9 in the United States. None arrived until February 1918.

At best, planning for the production of observation aircraft in the United States was confused in the fall of 1917. Just how confused the whole matter was is amply illustrated by the following officially approved programs, each superseding the one before it.[26]

COMPARISON OF DH-4 AND DH-9 PROGRAMS
SHOWING CHANGES IN GOALS FOR PRODUCTION

| DH-4 Aircraft | | | DH-9 Aircraft | | |
|---|---|---|---|---|---|
| 2 Aug. | 1917 | 7,000 | | | |
| 22 Aug. | 1917 | 5,000 | | | |
| 25 Aug. | 1917 | 15,000 | | | |
| 31 Aug. | 1917 | 6,000 | | | |
| 4 Sept. | 1917 | 15,000 | 5 Sept. | 1917 | 2,000 |
| 17 Oct. | 1917 | 250 | 9 Oct. | 1917 | 4,000 |
| 29 Oct. | 1917 | 1,000 | 29 Oct. | 1917 | 7,000 |
| 11 Feb. | 1918 | 4,500 | 1 Feb. | 1918 | 5,400 |

The wide variations in the numbers of aircraft believed to be required as reflected in the successive programs above reveal the ab-

24. *Aircraft Production,* Hearings before subcommittee of Senate Military Affairs Committee, 65 Cong. 2 Sess., 15 July 1918, *2, 755*; and Hughes Report, *Automotive Industries; the Automobile, 39,* No. 18, 750–10, 11.

25. British DH-4 production in the last quarter of 1917 was 427, DH-9 production was 5; however, in the first quarter of 1918 DH-9 production was 331, almost equal to DH-4 production, which amounted to 352. In the second quarter of 1918 only 70 DH-4's were produced in the United Kingdom against 1,156 DH-9's. Jones, *The War in the Air,* Vol. *3,* App. VII.

26. Hughes Report, p. 750–13. Not all fluctuations in the program are shown.

sence of a single comprehensive system of control over quantity. Likewise, the approved programs give no indication of control over quality, that is, production of the superior design (DH-9) to supersede its forerunner (DH-4) when possible. Expedience rather than planning seems to have been the order of the day.

In February 1918 officials in charge of production for the Equipment Division of the Signal Corps canceled the entire DH-9 program to concentrate upon the DH-4. Undoubtedly the delayed arrival of the DH-9 from England contributed to this decision. That no tactical aircraft of any type had been turned out in quantity in the United States after 10 months of war was probably also a factor. The real significance of the decision to produce the DH-4 rather than the DH-9 lay in that it was made by officials in charge of production without reference to the tactical elements of the service in combat.[27]

The DH-9 contracts were converted to cover production of the DH-4 instead. Two manufacturers, the Dayton-Wright Airplane Company and the Fisher Body Corporation, undertook to manufacture the aircraft in mass production. The manufacturers were successful in converting the DeHaviland design to carry a standard Liberty engine, and a flight test of the converted model was first made in October 1917. But the initial promise of the converted DH-4 was illusory. Although the redesign of the engine worked out satisfactorily, accessory equipment proved troublesome. The original sample DH-4 received in the United States had been designed for British armament and accessory equipment. In the mounting of machine guns manufactured in the United States, a number of small but vexing problems involving redesign of the cowling were encountered. Similarly, it was discovered that the synchronizing or interrupter gear for the machine guns had not been manufactured to fit the most recent model of the Liberty engine. This involved time-consuming rework during assembly. The same story was repeated in the instance of the special generator to supply power for the pilot's heated suit. Each problem retarded the beginning of actual mass production.

The charge that the Liberty engine delayed the program of construction was not substantiated. In May 1918 when the manufacture of the DH-4 in quantity finally did begin, the rate of production on Liberty engines was well in advance of the assembly of aircraft. The following table shows that the DH-4 ultimately did reach mass production.[28]

27. *Ibid.*, p. 750–15.
28. Mixter and Emmons, *Aircraft Production Facts*, p. 48.

PRODUCTION OF THE DH-4 IN THE UNITED STATES

| | | | |
|---|---|---|---|
| Feb. | 1918 | 9 | |
| March | 1918 | 4 | |
| April | 1918 | 15 | (first production models) |
| May | 1918 | 153 | |
| June | 1918 | 336 | |
| July | 1918 | 484 | |
| Aug. | 1918 | 224 | (including 100 sets of spares) |
| Sept. | 1918 | 757 | (including 100 sets of spares) |
| Oct. | 1918 | 1,097 | |
| Nov. | 1918 | 1,072 | |

Some of the first DH-4 aircraft were shipped to France where they arrived in May 1918. Sizable shipments, though, did not begin until the following month, and it was not until August that the first squadron of aircraft built in the United States actually crossed the enemy's lines. The reasons behind this protracted delay were manifold. The first group of aircraft to arrive was not equipped with armament. Facilities had to be made ready and an organization evolved to mount the armament before the Air Service of the AEF could become a fighting force. But more important than the problems of assembly in the field were the shortcomings intrinsic in the DH-4 itself.[29]

When the British DH-4 first began to appear in numbers on the front during the spring of 1917, its performance surpassed that of aircraft then in use. When the first squadron of the Air Service, AEF, went into combat in 1918, the DH-4 had itself been surpassed long since by aircraft of superior performance. Officers in the Air Service cabled from France that the DH-4 would not be acceptable for use in combat without drastic modification. Nevertheless, in the United States production was rolling, and shiploads of DH-4's continued to reach France over the protests of the AEF.[30] By the late summer of 1918, opinions as to the roles of production and operations were almost diametrically opposite to those held several months earlier. Then, the decisions made by the officials controlling production appeared to favor performance or quality while tactical units in the field demanded quantity. When at last aircraft in numbers did begin to arrive, the emphasis of the protest from the field shifted sharply to performance.

The charge of one engineer that the Bureau of Aircraft Production "railroaded" the DH-4 into production rather than the

29. "Final Report of the Chief of Air Service, AEF," *Air Service Information Circular, 2,* No. 180 (15 Feb. 1921), 21, 26, 40, 70.
30. *Aircraft Production in the United States,* Senate Report No. 555, p. 8.

DH-9 may have been only a half-truth.[31] It is apparent, however, that the necessity for making a showing exerted a tremendous pressure upon officials in charge of production.[32] The fault, of course, could never be laid entirely to these authorities. Some of the blame lay with the AEF. In June 1918 technical decisions on modifications in the DH-4 were going all the way up to the Chief of the Air Service, AEF, for approval in the absence of a properly constituted agency to deal with the problem.[33] Suggestions for modifying the DH-4 were thus slow in reaching the United States. There were, to be sure, officials who testified after the war that they knew all along that the DH-4 was inferior to the DH-9, but, significantly, they did little about it when the critical decision was made.

Colonel E. S. Gorrell, one of the two army aeronautical engineers with the Bolling Mission, declared that the DH-4 had been from the very beginning a stopgap for the DH-9.[34] The other army engineer with the mission, Lt. Col. V. E. Clark, asserted that the DH-4 had been put into production "simply because it was necessary to make an immediate showing," even though the authorities were "fully conscious" that the DH-9 was a better aircraft.[35] It might well be argued that these two officers were pleading defensively, trying to explain their failure to emphasize the stopgap nature of the DH-4 at the time of the original or Bolling report which shaped policy. The Chief of the Air Service, AEF, was not so suspect. As he was an outsider who took command in May 1918 from an entirely different arm of the service, his testimony that the DH-4 was obsolete and used only until the DH-9 could be put into production was probably free from defensive bias.[36] As late

31. Notes on conversation with Lt. Col. V. E. Clark by R. M. McFarland at McCook Field, Dayton, Ohio, 2 March 1919, NA, BAP Hist. Box 20, 452.1 Airplanes, General.

32. P. C. March in *The Nation at War* (Garden City, N.Y., Doubleday, Doran, 1932), p. 207, claims that Assistant Secretary of War J. D. Ryan's contribution to production was to promise that there would be "no more changes authorized," thus implying that the seat of the trouble was the AEF. March's contention would, by inference, place Ryan in the light of utterly failing to grasp the importance of performance *vs.* numbers, a charge not substantiated.

33. Memo, Maj. H. A. Toulmin to Brig. Gen. M. M. Patrick, 8 June 1918, cited in Toulmin, *Air Service, AEF*, pp. 87–8.

34. Testimony of Col. E. S. Gorrell, *War Expenditures,* House Hearings, 66 Cong. 1 Sess., Serial 2, Pt. 6, 4 Aug. 1919, p. 211.

35. Notes on conversation with Col. V. E. Clark by R. M. McFarland, 2 March 1919, NA, BAP Hist. Box 20, 452.1 Airplanes, General.

36. Testimony of Col. M. M. Patrick, *War Expenditures,* House Hearings, 66 Cong. 1 Sess., Serial 2, Pt. 4, 4 Aug. 1919, pp. 221–2. See also Testimony of Maj.

as August 1918 the DH-9 was still in the stage of flight testing in the United States. The decision to go into production was yet unmade.[37]

In the final analysis, the measure of achievement of the aerial weapon must always be determined by the number of superior aircraft available for use in combat and actually on the front. When the war ended, the Air Service, AEF, consisted of 20 pursuit squadrons, 24 observation squadrons, some of which were serving as day-bombers, and one lone squadron of long-range night or strategic bombers.[38] At first glance this array is more impressive than the facts warrant. Of the total number of aircraft received by the AEF up to the Armistice, 6,287 in all, a large proportion were of foreign origin. The following table indicates the comparatively small contribution from production in the United States.

| | |
|---|---|
| From French sources | 4,791 |
| From British sources | 261 |
| From Italian sources | 19 |
| From U.S. sources | 1,216 |

Of the aircraft contributed by the United States, all but three odd experimental items were DH-4's. Only 960 of these were actually at the front; the remainder were being used at training bases in the theater.[39]

---

B. D. Foulois, *War Expenditures,* House Hearings, 66 Cong. 1 Sess., Serial 2, Pt. 6, 6 Aug. 1919, pp. 367–8.

37. *Aircraft Production in the United States,* Senate Report No. 555, p. 9. Some 6,000 DH-9's (called US-9) were placed on contract, but none was completed by the war's end. See *Report on Aircraft Surveys . . . ,* House Document No. 621, p. 7. Compare this with the 1,156 British DH-9's produced in the second quarter of 1918. See Jones, *The War in the Air,* Vol. *3,* App. VII.

38. To avoid confusion the DH-4 has here been consistently labeled as an observation aircraft. Army-cooperation or service aircraft might be more exact definitions, but the terms are less well known. In general, the DH-4 was used for observation, infantry contact-patrol, artillery fire-control, and short-range or tactical day-bombing. The type lacked the speed and other performance characteristics of pursuit or fighter aircraft and the range and bomb-carrying capacity of the strategic night-bomber. "Progress Report of Air Service Activities, AEF, as of 11 Nov. 1918," p. 1, 30 Nov. 1918, Air Service Exec. Sect., AEF, National War College Library, UG576.3.

39. E. S. Gorrell, *The Measure of America's World War Aeronautical Effort* (Northfield, Vt., Norwich University, 1940), p. 35. Gorrell's figures, based on data compiled by the Air Service, AEF, vary slightly from those of Mixter and Emmons, *Aircraft Production Facts,* p. 58, quoting from General Staff Statistics Branch, Report No. 91. Gorrell also differs from the Secretary of War, *Annual Reports of the War Department,* 1918, *1,* 52–3. Gorrell's figures are probably more nearly correct. However, Jones, *The War in the Air, 6,* 81, puts the British contribution at 320. This figure may include items sent to the United States as well as to the AEF.

A summary of the development of military aircraft in the United States is scarcely praiseworthy. In November 1918 there were no fighter aircraft in the Air Service, AEF, save those procured from foreign sources.[40] There was a relatively large number of observation aircraft at the front, but these were of inferior performance and dangerously obsolete. The officials responsible for developing the aerial weapon of the United States had failed to achieve the necessary balance between superior performance and mass production. In their efforts to perfect a superior fighter they produced none at all.

40. The performance of the French-supplied fighter aircraft may not have been so inferior as charged by a witness before the Senate Military Affairs Committee who described the planes as "antiquated" and already "discarded" by the French themselves. But it is safe to assume that the foreign aircraft procured by the U.S. were not exclusively the best produced. See *Aircraft Production in the United States*, Senate Report No. 555, p. 10.

## Chapter *VIII. The Development of Air Weapons: Bombers*

THE DOCTRINE OF aerial warfare, or the approved concept of the military functions to be carried out by the aircraft sent to France, imposed few difficulties in the case of aircraft for observation and pursuit. Both functions were almost universally accepted in military circles, the one to serve as "the eyes of the army," the other to defend the "eyes" from enemy attack and prevent enemy observers from performing the same operation. On the other hand, doctrine regarding the employment of bombing aircraft had never been clearly defined or generally accepted in military circles. As a consequence, the uncertain status of doctrine on bombardment exerted an overwhelming influence upon the production of bombers.

The military organization for providing aerial weapons had shown itself deficient in supplying aircraft where doctrine constituted no real problem. Where doctrine was in doubt, the military organization for supplying aircraft was to prove even more deficient. To help in measuring the achievement of that organization in producing bombers during World War I it will be convenient to review briefly the evolution of bombardment aviation. After the outbreak of war in Europe during August 1914 and before the United States entered the war in April 1917, aerial doctrine underwent a revolutionary development in the armies of the warring Powers. Unfortunately, the security measures of the combatants coupled with an inadequate system of liaison left the United States in a situation of technical as well as political isolation. While the concept of aerial warfare leaped ahead in Europe, the United States followed gropingly. Doctrine in the Signal Corps was the product of two factors: extremely limited operational experience with a handful of training aircraft and domestic interpretations of the scanty reports of military attachés.

By 1915 the war in Europe made it clear that aircraft could be employed to perform a number of functions including reconnaissance, prevention of enemy reconnaissance, artillery fire-control or

adjustment, transportation, and the destruction of enemy materiel
and personnel. Not all of these missions were considered of equal importance by military officers in the United States. The Chief Signal Officer, who was directly responsible for the development of the
air weapon, reported in 1915 that "the useful, approved, and most
important work of aircraft" was to be found "chiefly in reconnaissance." [1] He dismissed the defensive and transport roles of aircraft in a single sentence as "obvious." The question of bombardment and the offensive role of aircraft, broadly considered, he
treated circumspectly. "Whatever may be the opinion of military
men as regards the offensive importance of aircraft . . . there is
no longer a question as to the value of the aeroplane in . . . reconnaissance work . . ." [2] This expression of doctrine reflected the
opinion of the Signal Corps until April 1917.

When the United States entered the European war there may
have been no fixed opinion throughout the army regarding the relative importance of observation, pursuit, and bombing aircraft.
Nevertheless, in the Signal Corps there was a marked predisposition to inflate the function of observation at the expense of all
others. And the Signal Corps was in a position to determine almost entirely the nature of the aerial force to be developed. The
reader will recall from Chapter 3 that the arrival of a cable from
Premier Ribot in France suddenly precipitated an aircraft program for the United States. The program established a target for
production expressed entirely in terms of numbers. The cable, as
received, failed to include any information regarding the proportion of functional types of aircraft to be provided. As a consequence, determination of the relative weight of squadrons for
pursuit, observation, and bombardment in the new air force fell
entirely to the initiative of officials in the United States. Since the
prevailing doctrine of the Signal Corps regarding aerial warfare
favored reconnaissance, it was only natural that these officials allocated to aircraft for observation and to pursuits for the defense
of observation a dominant proportion or nearly 89 per cent of the
total program. [3]

When the Bolling Mission went to Europe, the members of
the group were greatly impressed with the potentialities of bom-

1. *The Service of Information, United States Army,* Office of Chief Signal Officer,
Circular No. 8, 1915, p. 23.
2. *Ibid.,* pp. 21–3.
3. Joint Army-Navy Technical Board to Secretary of War, 29 May 1917, NA,
BAP Misc. Hist. Box 1, 321.9 A.S., Training. The JANTB proposal specified 3,000
observation, 5,000 fighter, and 1,000 bombing aircraft in the tactical force.

bardment. The official report of the mission in August 1917 favored a policy specifying, first, training aircraft; second, aircraft for close-support or tactical cooperation with the ground forces; and third, a strategic offensive force made up of all excess over the first two categories. In the early fall of 1917 Colonel Bolling noted that it was the "settled conviction" of the mission that the importance of "bombing operations with direct military ends in view" could not be exaggerated.[4] Lieutenant Colonel V. E. Clark took an even stronger position, declaring that intensified night bombing would "put an end to the war far more quickly than sending one or two million men to line the trenches." Major E. S. Gorrell was even more specific in assessing the role of bombardment. He felt that the Air Service, AEF, would be certain to wreak "immense destruction" upon German morale and materiel if it could place in the field a sufficiently large number of night-bombers to carry out a "systematic bombardment" of Germany. Major Gorrell left little doubt as to the importance he attached to bombardment:

> This is not a phantom nor a dream, but is a huge reality capable of being carried out with success if the United States will only carry on a sufficiently large campaign for next year, and manufacture the types of airplanes that lend themselves to this campaign, instead of building pursuit planes already out of date here in Europe.

He considered the 1,000 bombers planned in the official program in the United States a sadly deficient number. A force ranging from 3,000 to 6,000 bombers, he believed, would be far more adequate.[5]

A bombing force of 6,000 aircraft would have represented a complete reversal of official policy. Where the program of production approved by the General Staff stipulated aircraft for pursuit, observation, and bombardment in the ratio of 5:3:1, Major Gorrell's proposal would substitute a ratio of 5:3:6, giving a preponderant weight to the offensive character of the air arm. Gor-

4. Bolling message of 4 Sept. 1917 quoted in A. B. Gregg, "History of the Caproni Biplane," 1919, NA, BAP Hist. Box 21, 452.1 Caproni History; and Memo, Lt. Col. V. E. Clark to Chief Signal Officer, 12 Sept. 1917, reprinted in *Aircraft Production*, Hearings before subcommittee of Senate Military Affairs Committee, 65 Cong. 2 Sess., 15 July 1918, *2*, 799.

5. Copy, memo, Maj. E. S. Gorrell to Col. R. C. Bolling, 15 Oct. 1917, and Cable No. 96, Bolling to Chief Signal Officer, 13 Aug. 1917, NA, BAP Hist. Box 21, 452.1 Caproni Contract. See also Memo, Lt. Col. V. E. Clark to Chief Signal Officer, 12 Sept. 1917, in *Aircraft Production*, Hearings . . . Senate Military Affairs Committee, 65 Cong. 2 Sess., 15 July 1918, *2*, 799.

rell's recommendations, like those of the other members of the Bolling Mission, may have been factors influencing official policy, but they did not become official policy. For that matter, neither did the approved program of the General Staff long remain the accepted plan. By the middle of October 1917 the Air Service, AEF, had drafted a program which determined the character of the air arm along somewhat different lines from those in the original program prepared in Washington. Where the original or approved program of the General Staff had proposed a force in which aircraft for pursuit, observation, and bombardment were to be supplied in the ratio of 5:3:1, the program of the AEF specified a ratio of 3:2:1½.

The approved program of the Air Service, AEF, detailing the composition of the air arm by relative proportion of functional types, underwent a large number of transformations and permutations before aircraft were actually received from the United States. In January 1918 officials of the AEF appeared to accept in principle the original program of the General Staff. In place of the earlier ratio, approved by the AEF, of 5:3:1 for pursuit, observation, and bombardment, the revised program of the AEF substituted the nearly comparable ratio, 6:2:1. This was not to remain the accepted expression of doctrine for long. After a number of alterations in the official program the Air Service, AEF, issued a plan in June 1918 which was radically different from all those previously prepared. This new program increased the offensive force at the expense of aircraft for observation, making a new ratio in pursuit, observation, and bombardment of 3:1:2½ in the air arm.[6]

The violent changes in the AEF's programs which occurred between October 1917 and June 1918 obviously did not stem from experience in combat. The AEF had in France no more than a handful of American-built aircraft by the beginning of June 1918. At that time the Air Service consisted of but five squadrons equipped with aircraft of foreign manufacture.[7] Since the five squadrons did not include a single one for bombardment, it is apparent that Allied operations outside the AEF accounted for the

6. See above, Chapter 3, and "Formulation and Distribution of Programs," Table XII, Sept. 1919, NA, BAP Hist. Box 9, 334.8 Overseas Missions.

7. E. S. Gorrell, *The Measure of America's World War Aeronautical Effort* (Northfield, Vt., Norwich University, 1940), p. 30. See also "Final Report of the Chief of Air Service, AEF," *Air Service Information Circular, 2,* No. 180 (15 Feb. 1921), 5.

doctrinal changes implied by the successive official programs for aviation. The evolution of British doctrine appears to have influenced thinking in the AEF.

In the summer of 1916 British units based near Belfort conducted strategic bombing attacks against the German industrial area of the Saar. These limited operations were curtailed early in 1917 when Sir Douglas Haig, commander-in-chief of the British troops in France, requested the services of the units for close-support with the British ground forces.[8] But just about the time this action took place, German air raids on London brought the whole question of strategic doctrine to a head.

The Chief of the Imperial General Staff threw the question of the aerial weapon into sharp focus when he proposed an increase in the strength of British aviation, even at the expense of other weapons. Any such expansion would inevitably lead to conflict with the existing services. The War Cabinet, with remarkable clarity of purpose, established a Priorities Committee giving aviation a voice in an echelon comparable to those of the army and navy. This preliminary step in the fall of 1917 led to the formation of the Air Ministry at the end of the year.[9] By creating the Air Ministry the British government established an organization with effective authority to execute a strategic doctrine. This doctrine had been formulated partly by projection of limited British operational experience and partly as a result of German proddings in the form of demoralizing raids over London. Sir William Weir, Secretary of State for the newly formed Royal Air Force, summed up the doctrine of the Ministry succinctly in pointing out that "continuous bombing of German industrial centers" presented "very important possibilities" and that valuable results might be obtained even with the small forces available. The Ministry contended that the German economy was a profitable target fully justifying diversions of strength from the ground armies.[10]

Marshal Foch, as commander-in-chief of the Allied armies, opposed the concept of the British Air Ministry with the traditional point of view of the ground forces. The enemy's army, he declared, was the enemy's strength. Only by a concentration of force to

8. H. A. Jones, *The War in the Air* (5 vols., *2–6;* Oxford, Clarendon Press, 1922–37), *6,* 121.

9. *Ibid.,* pp. 2–17.

10. Memo, Sir William Wier, Secretary of State for the Royal Air Force, on the Responsibility and Conduct of the Air Ministry, 23 May 1918, cited in full in Jones, *The War in the Air,* Appendix Volume, App. VII.

destroy that army would it be possible to attain victory. Bombers, Foch said, should be used for strategic raids against the enemy's economy only as a secondary function.[11]

His conception of the primary objective of bombers did not blind Foch to their potentialities. He drew up an elaborate plan of attack on the German economy, but his plan stipulated the use of bombers only after the requirements of the armies in the field had been met or during lulls between battles.[12] So the military representatives of the Allied Supreme War Council formulated a program of bombardment during August 1918 to be carried out by an Inter-Allied Bombing Force within the frame of reference laid down by Foch. Brevet General Tasker H. Bliss, military representative for the United States to the Allied Supreme War Council, joined the representatives of the other Powers in signing the joint note forming the Inter-Allied Bombing Force.[13] The end of the war came before the Inter-Allied Force materialized. Nevertheless, the intention is evident. The Allies and the United States were moving in the direction of strategic bombardment. Even though conceived within the orbit of Foch's command and compelled by that circumstance to consider attacks on the German economy as secondary, the proposed force was strategic in its potential.

As the cumbersome machinery of Allied control evolved this plan, the British demonstrated what could be done in practice. In June 1918 the British Air Ministry actually organized a strategic bombing arm known as the Independent Force, although, to be sure, the force consisted of but five squadrons, only two of which were composed of long-range night-bombers capable of deep penetrations into enemy territory. Throughout the summer of 1918 this force repeatedly attacked German industrial centers and rail junctions. Handley-Page bombers carrying individual bombs weighing 1,650 pounds were used during October, giving a foretaste of what was to come. Despite the future possibilities embodied in explosives of this weight, the Independent Force was never more than a shadow of its planned strength. When the Armistice came, the British Independent Force consisted of only nine squadrons, and some of these were equipped not with bombers but with aircraft converted from types designed for reconnaissance which had

11. *Ibid.*, App. VIII, Memo, signed by Foch, on the subject of an Independent Air Force, 14 Sept. 1918.

12. *Ibid.*, App. X, Memo, Foch to Clemenceau, 13 Sept. 1918.

13. *Ibid.*, App. IX, Joint Note No. 35, Military Representatives of the Supreme War Council to Supreme War Council, 3 Aug. 1918.

been considered obsolete since 1916. Nonetheless, small as this force was, it represented an organizational framework upon which the Air Ministry intended to carry out a program of strategic bombardment on the German economy. Only the pace of production and development of design held this policy in check.

At the Armistice the output of British aircraft was rapidly overtaking the Ministry's program, and development of design had opened new vistas to strategic bombardment. As early as July 1917 the British had gone into production on the long-range, twin-engine Handley-Page night-bomber. By September some 400 had been placed on contract. But of greater importance, experimental contracts were placed during the summer of 1917 for even larger Handley-Page superbombers, four-engine aircraft capable of carrying 7,500 pounds of bombs to Berlin. Three of these machines, completely superseding the twin-engine Handley-Page, were available at the end of the war. Some 250 were on contract.[14]

Even a brief excursion into the tale behind the growth of British doctrine on strategic bombardment is sufficient to suggest the character of the influences at work upon the shaping of policy in the Air Service, AEF. The programs promulgated for aviation by the AEF during the fall of 1917, before the formation of the Air Ministry, reflected the influence of the ground force. During the spring of 1918, after the creation of the Air Ministry, they reflected an increasing acceptance of the idea of strategic bombardment. In fact, the program of June 1918, the month in which the British Independent Force appeared, almost reversed aerial doctrine of the AEF by stipulating an increase in bombers to more than twice the strength in aircraft for observation.

Appreciation of the doctrine of bombardment was by no means limited to military circles overseas. During June 1918, shortly after the British Independent Force had been organized, the Director of Military Aeronautics in the United States prepared a study on strategic bombardment for the Chief of Staff, Gen. P. C. March. Coming from an authoritative spokesman responsible for the formulation of requirements, the director's study represented, by implication, a significant indication of the trend in aerial doctrine. The burden of the study was expressed in one paragraph:

> Apart from the production of aircraft sufficient for the tactical and strategic needs of the Air Service, AEF, consideration must be given the production of a large number,

14. Jones, *The War in the Air, 6,* 135, 150, 164, 168, 173–4.

—probably in excess of 1500—airplanes for a "long dis-
tance independent bombing force," to borrow a phrase
from the British . . . to operate on a bombing campaign
against German industrial centers in cooperation with British and French forces of a similar nature.

To this end the director urged beginning immediately on production of superbombers, preferably the Handley-Page four-engine
model.[15] The Chief of Staff must have been impressed with the
arguments raised by the Director of Military Aeronautics, for
on the day the director's study reached him he wrote to the Bureau
of Aircraft Production, pointing out that strategic conditions demanded the production of aircraft with a radius of 1,500 miles and
a bomb load of approximately 4,500 pounds. These superbombers,
the Chief of Staff indicated, would form a "long-distance independent bombing force." He requested a report on the possibility
of producing 570 such bombers without interfering with the production of aircraft for the tactical and limited strategic needs of
the Air Service, AEF.[16]

The sum total of the evidence available shows that in June 1918,
both in the AEF and in the United States, the doctrine of strategic
bombardment was finding increasing acceptance and approval.
Then, at the end of July, the Chief of the Air Service, AEF,
promulgated a revised program. This version, the final one of the
war as it ultimately turned out, was known as "the 202 program."
Of the 202 squadrons planned for the AEF 60 were to be pursuit,
101 were to be observation, and only 41 were to be squadrons of
bombers. The new allocation of squadrons by functional types
represented a ratio for pursuit, observation, and bombardment of
approximately 3:5:2. It will be recalled that the previously approved program of the AEF had established a 3:1:2½ ratio.[17]

15. Maj. Gen. W. L. Kenly, Director of Military Aeronautics, to Gen. P. C. March,
Chief of Staff, 29 June 1918, NA, BAP Misc. Hist. Box 1, 321.9 Organization of
DMA.

16. Copy, Chief of Staff to Director Aircraft Production, BAP, 29 June 1918, and
Copy of reply, 10 July 1918, NA, BAP Hist. Box 23, 452.1 Handley-Page. The
former document is particularly interesting in that it appears to conflict with
March's contention in *The Nation at War* (Garden City, Doubleday, Doran, 1932),
pp. 208–9, that he "strongly urged" the Secretary of War not to approve a joint
bombing force under British leadership since such action would take Pershing's air
force from him. March may have objected to British control more than he did to
the idea of an independent strategic force. The document cited above indicates that
he did nevertheless favor a bombing force *apart* from Pershing's army units.

17. The disparity between the programs of June and July is even more strikingly
revealed by a comparison using numbers of squadrons instead of ratios of squadrons:

Did this drastic change in the proposed composition of the Air Service, AEF, mark a reversal of policy from emphasis on bombers to emphasis on observation? Was this change in composition induced by a new doctrine regarding the best operational use of the aerial weapon? Or did the 202 program merely reflect a realistic consideration of the possibilities of production? A review of the production of bombers in the United States during 1918 may help to answer these questions.

The formal report of the Bolling Mission at the end of July 1917 recommended, the reader may recall, the Italian three-engine Caproni as a suitable bomber for production in the United States. When the mission made this decision the Caproni had already proved itself in operations along the front, but during the summer of 1917 a competitor, the British twin-engine Handley-Page bomber, began to appear in increasing quantities. When Lieutenant Colonel Clark returned to the United States from his service with the mission, he recommended the Caproni bomber and mentioned the Handley-Page as an acceptable substitute.[18] The requirements of simplicity and standardization for production made it evident that one or the other but not both should be selected for manufacture. Choosing between the two bombers proved extremely difficult. In October Colonel Bolling cabled instructions to concentrate on the Caproni. In the following month a cable arrived in Washington over Pershing's signature which favored maximum production of the Handley-Page but advised withholding consideration of the Caproni.[19] The Joint Army-Navy Technical Board straddled the issue and recommended both for production in a revised program drawn up in November 1917. Even though

|  | Pursuit | Observation | Bomber | Total Squadrons |
|---|---|---|---|---|
| June program | 120 | 40 | 101 | 261 |
| July program | 60 | 101 | 41 | 202 |

It will be noticed that the strength in squadrons for pursuit was cut in half in the program of July and the strengths in squadrons for observation and bombing were almost exactly reversed. See "Formulation and Distribution of Programs," Table XII, Sept. 1919, NA, BAP Hist. Box 9, 334.8 Overseas Missions.

18. Copy, memo, Lt. Col. V. E. Clark to Chief Signal Officer, 12 Sept. 1917, NA, BAP Hist. Box 21, 452.1 Caproni Data. This report summarizes the doctrine of strategic bombardment effectively. See above, p. 135 n. 4, for reference to published version.

19. Copy, cable, Col. R. C. Bolling to Asst. Secretary of State W. Phillips, 9 Oct. 1917, and Copy, Pershing cable of 27 Nov. 1917, NA, BAP Hist. Box 23, 452.1 Handley-Page. The Bolling cable, which arrived via his brother-in-law Phillips in the State Department, is but another example of the confusion of channels for technical information coming from the AEF.

production of two such types would cause a great many problems, not only of supply but in manufacture, both the Secretary of War and the Secretary of Navy approved this program.[20]

By February 1918 the Aircraft Production Board was understandably disturbed at the necessity for duplicating production at a time when factories in the United States were already overburdened with imperative orders. The authorities in charge of production urged the Joint Army-Navy Technical Board to come to a decision. Unfortunately the joint board was scarcely in a position to make such a decision since the information from abroad was inadequate and conflicting.[21] In the absence of technical information upon which to base decisions, planning for production in the United States was erratic, to say the least. Just how erratic this planning could be is indicated in the following table showing the fluctuations in the targets set for production of the Caproni.[22]

| Date Program Formulated | Number of Bombers Planned |
|---|---|
| 9 Aug. 1917 | 500 |
| 16 Aug. 1917 | 9,000 |
| 22 Aug. 1917 | 2,000 |
| 24 Aug. 1917 | 500 |
| 28 Sept. 1917 | 1,000 |
| 19 Feb. 1918 | 50 |
| 3 May 1918 | 250 |

It will be noticed that the low point in production planned for the Caproni bomber coincides with the period at which the AEF demanded a ratio in squadrons for pursuit, observation, and bombardment of 6:2:1, as previously mentioned. The influence of AEF plans for squadrons upon the programs for production in the United States was not always quantitatively commensurate, but it was nevertheless very real.[23]

The original intention of the Bolling Mission had been to secure Caproni bombers in Europe until production could begin in the United States. To this end the Bolling Mission began negotia-

20. Hughes Report, 25 Oct. 1918, *Automotive Industries; The Automobile, 39,* No. 18 (31 Oct. 1918), 750–14.

21. While the board tried to reach a decision it was reported that production of the Handley-Page in England was even then one per week. Copy, minutes of Aircraft Board Meeting, 8 Feb. 1918, NA, BAP Hist. Box 21, 452.1 Caproni Contract. This report, however, was false, as shown by postwar British records. Jones, *The War in the Air,* Vol. *3,* App. VII.

22. Hughes Report, *Automotive Industries; The Automobile, 39,* No. 18, 750–13. Not all minor changes are shown.

23. See, for example, Copy, resolution of Aircraft Production Board, 30 Oct. 1917, NA, BAP Hist. Box 21, 452.1 Caproni Contract.

tions to establish a Caproni factory at Bordeaux. The project was abandoned during February 1918, according to the Chief of the Air Service, "in view of the . . . decision against an increase in the number of bombing squadrons." [24] Whatever Caproni bombers could not be borrowed from Allied air forces would have to come from the United States.

The prospects for domestic production of bombers were not bright. Although the Caproni had been selected by the Bolling Mission at the end of July 1917, in November detailed drawings of the bomber had not yet reached the United States.[25] During January 1918, nearly six months after the initial selection, a technical mission arrived from Italy to facilitate Caproni production. This group, consisting of the colorful Capt. Ugo d'Annunzio, several experienced bomber pilots, and some skilled artisans from Italian aircraft factories, set to work converting the Caproni for use with the Liberty engine. Progress was far from satisfactory. Italian shop practice differed markedly from that in the United States, and the drawings proved difficult to convert to local usage. These engineering troubles were aggravated by more serious external problems. The Caproni technicians were commercially rather than diplomatically accredited, and a conflict soon developed between d'Annunzio and the Italian Embassy as to the authority of the group. Engineering waited on protocol.[26]

The pace of production of Capronis was seriously retarded by the frequent shifts in the policy of the War Department, shifts in prime contractors as well as in quantity. During September 1917 the Aircraft Production Board considered giving a contract for 500 Capronis to the Curtiss Aeroplane and Motor Corporation. In February 1918 this project gave way to a contract for 50 Capronis with the Standard Aircraft Corporation. The Fisher Body Corporation received a verbal contract for 500 bombers in April, and in June formal contracts were issued to Fisher Body and Curtiss for 500 Caproni bombers each. Not until July 1918 did the first Caproni built in the United States enter upon flight tests, and the

---

24. Cable from Brig. Gen. B. D. Foulois, Chief of Air Service, 18 Feb. 1918, cited in copy, minutes of Aircraft Board, 19 Feb. 1918, NA, BAP Hist. Box 21, 452.1 Caproni Contract.

25. Cable, Maj. L. S. Horner to Col. R. C. Bolling, 21 Nov. 1917, NA, BAP Hist. Box 21, 452.1 Caproni Contract.

26. Frank Brisco, "Data on the History of the Caproni Project," 14 May 1919, NA, BAP Hist. Box 21, 452.1 Caproni Contract. Brisco was one of the officers participating in the Italian Caproni project. After the war he compiled a brief account of his role for the Air Service.

model was still in the experimental stage when the Armistice came.[27]

The record of difficulties encountered in putting the Caproni into production shows plainly why the authorities finally determined to concentrate their efforts upon the Handley-Page. The British bomber was decidedly inferior to the Italian aircraft, both in range and in bomb load; without guns, ammunition, or bombs the Handley-Page developed a high speed of only 97 mph and a 14,000-foot ceiling.[28] Despite this inferior performance the authorities favored the Handley-Page over the Caproni bomber because they were convinced that many months must elapse before the Caproni could approach mass production. Once again a decision on production superseded a requirement for combat. The Handley-Page was itself no insignificant problem in production. The bomber comprised a total of more than 100,000 parts, a formidable manufacturing project under the best of conditions. In 1918 it was impossible to fly the Atlantic, and inasmuch as the aircraft had a 100-foot wingspan, the unavoidable alternative was to ship the bombers dismantled to Europe and assemble them there. During January 1918 the Chief of the Air Service, AEF, reached an agreement with British authorities whereby the United States would supply sets of parts for Handley-Page bombers and the British would assemble the parts in an idle Lancashire cotton factory.[29] Unfortunately few sets of parts ever reached England.

The first "complete" set of drawings for the Handley-Page bomber reached the United States during August 1917, giving the British bomber a decided advantage over the Italian Caproni. But the forehandedness of the British in sending drawings to the United States promptly after the Bolling visit had its disadvantages. Just as in the case of the DH-4 where subsequent British changes in design produced the DH-9 of superior performance, so, too, subsequent changes in design in the Handley-Page im-

27. A. B. Gregg, "History of the Caproni Biplane," 1919, NA, BAP Hist. Box 21, 452.1 Caproni History. See also Col. G. W. Mixter and Lt. H. H. Emmons, *United States Army Aircraft Production Facts* (Washington, GPO, 1919), p. 53; *Report on Aircraft Surveys* . . . , House Document No. 621, 66 Cong. 2 Sess., 19 Jan. 1920, p. 4; Hughes Report, *Automotive Industries, The Automobile, 39,* No. 18, 750–15; and *Aircraft Production,* Hearings before subcommittee of Senate Military Affairs Committee, 65 Cong. 2 Sess., June to Aug. 1918, Vols. *1* and *2 passim.*

28. Figures cited are for performance of Handley-Page bomber built in the United States.

29. The agreement of Jan. 1918 is sometimes known as the Rothermere-Foulois Agreement; see H. A. Toulmin Jr., *Air Service, American Expeditionary Force, 1918* (New York, D. Van Nostrand, 1927), pp. 327–8.

proved its performance. But each change caused consternation in the United States. Twice during the winter of 1917–18 new sets of drawings arrived from England. Each time large numbers of parts already in process of manufacture had to be reworked with a consequent delay in completion. Engineering difficulties encountered in converting the airframe of British design for use with the Liberty engine caused the usual amount of trouble. Moreover, the Handley-Page suffered from the same contractual instability and uncertainty which marked the course of Caproni production in the United States.[30]

When the war ended, the Air Service, AEF, had two squadrons of Handley-Page bombers in training in England. Not a single night-bomber manufactured in the United States during World War I ever reached the front.[31] This unhappy fact may go far toward answering the question posed earlier in this chapter: did the program of July 1918 in which the AEF called for a pursuit, observation, and bomber ratio of 3:5:2 express declining confidence in bombardment, or was it a program based on the realities of production in the United States? The sorry record made in the production of bombers during the war gives every reason to support the latter view. Moreover, there is substantial evidence from another quarter to further this contention.

At about the same time that the Air Service, AEF, formulated the program for 202 squadrons in July 1918, the Chief of Staff, AEF, suggested a revised program. In this version the Air Service, AEF, would comprise a total of 358 squadrons with substantial increases in offensive force but no increase whatsoever in units for observation. When this plan is viewed comparatively with the program proposed by the Air Service in July there can be no question but that the AEF had not changed its doctrine of bombardment since the spring of 1918.[32]

| | *Pursuit* | *Observation* | *Bombers* | *Total* |
|---|---|---|---|---|
| Air Service Program of July 1918 | | 60 | 101 | 41 | 202 |
| Program Proposed by the Chief of Staff, AEF | 147 | 101 | 110 | 358 |

It may be argued that the former program represented realism in the Air Service more than it did the aerial doctrine of the AEF.

30. See above, p. 144 n. 27, and NA, BAP Hist. Box 23, 452.1 Handley-Page Data file.

31. "Final Report of the Chief of Air Service, AEF," *Air Service Information Circular, 2,* No. 180, 62–3.

32. "Formulation and Distribution of Programs," Table XII, and figures quoted from Chief of Staff, AEF, to Chief of Air Service, AEF, 29 July 1918, NA, BAP Hist. Box 9, 334.8 Overseas Missions.

There is still further evidence to indicate that the policy of the AEF embraced the doctrine of strategic bombardment despite the predominance of squadrons for observation in the 202 program. The action of Gen. Tasker Bliss who joined the military representatives of the Allies during August 1918 in proposing an Inter-Allied Bombing Force certainly sustains this view.[33] All of the evidence, weighed collectively, would appear to show that the program of July 1918 or so-called final program for 202 squadrons was not an expression of the official doctrine of the AEF; nor did it intentionally underemphasize the role of bombardment, as some postwar advocates of the Air Service were inclined to believe. The 202 program was apparently nothing more than an expedient plan based on calculations of probable rates of production.

When the war ended at the point where the program of the AEF had been adjusted to meet the probabilities of production rather than doctrine, the composition of the air force, as planned, became a dangerous and faulty precedent for postwar analysis. Just how the postwar planners made use of this precedent is a matter of extreme importance, for the lessons of World War I dominated the whole concept of air power for the 20 years of the long armistice. What the heritage of the United States in aerial weapons actually was at the end of the war remains for the following chapters to tell.

33. Jones, *The War in the Air,* Appendix Volume, App. IX, Joint Note No. 35, Military Representatives of the Supreme War Council to Supreme War Council, 3 Aug. 1918.

*PART THREE: Heritage of the Air
Weapon from World War I*

## Chapter IX. The Postwar Air Service and Development of the Air Weapon

WHAT DID THE United States learn about the air weapon from the experience of World War I? The military air arm of the nation grew from a mere handful of men with one or two aircraft to an organization comprising more than 150,000 men and 15,000 aircraft. Surely an undertaking of this magnitude could scarcely fail to yield a number of significant lessons for the future. Individual observers, officers and civilian officials alike, may well have recognized a great many of these lessons from wartime experience with aircraft. But experience of individuals is elusive. The effective heritage of the nation in terms of the aerial weapon can best be measured by the institutions and organizations, the procedures and policies, formally and authoritatively adopted to operate the air arm at the end of the war. The heritage from World War I was not to be found in the lessons learned but rather in the lessons which were not only learned but applied in shaping the air arm of the future.

The most tangible product of the war was the Air Service of the United States Army. With the Army Reorganization Act of June 1920 Congress established the Air Service as a combatant arm of the line, coordinate with the Infantry, Cavalry, and other arms. Congress authorized some 1,500 officers and 16,000 enlisted men to make up the new service with a major general as chief.[1] The air arm, as a result of the limited experience of World War I, had been moved from the status of a useful weapon in a subordinate position within one of the technical services to a self-sufficient role in the combat line.

There were other lessons learned and applied from World War I. Officials of the postwar Air Service believed the war had "clearly demonstrated" that the air arm could not function effectively while divided into two coequal but mutually independent halves as had been the case under the Bureau of Aircraft Production and the

1. 41 Stat. 768, Public No. 242, 4 June 1920.

Division of Military Aeronautics.[2] To remedy this evil the post-war Air Service specifically aligned both operations and supply under a single responsible chief. One might well argue that in so doing the War Department had shown itself exceptionally capable of absorbing the lessons of experience. Certainly the Air Service of 1920 constituted an important advance over the Aviation Section of the Signal Corps in 1917. The advances accorded the air arm by the War Department were undoubtedly relatively important, but did they represent any significant portion of the lessons available from the experience of the war?

The introductory chapter of this study suggested a hypothesis contending that superiority in weapons required an understanding of two basic considerations. The first of these involved a conscious recognition of the importance of formulating military doctrine to exploit to the utmost the superior weapons produced. At least one measure of the air arm's heritage from World War I would be to determine the extent to which the new Air Service recognized the importance of better weapons rather than more weapons and the importance of formulating doctrine to exploit those better weapons. A subsequent chapter will discuss the problem of doctrine; this chapter will consider only the extent to which the air arm learned the importance of quality over quantity or better weapons rather than more weapons. It will be remembered that during the Civil War the federal authorities favored the muzzle-loading musket over the repeating rifle, and at the end of the war national armories were heavily stocked with obsolete weapons. Did the War Department repeat this pattern in the case of the air weapon? After World War I did the Air Service favor a policy of more weapons or a policy of better weapons? A brief résumé of wartime practice and the impact of that experience upon post-war thinking should suffice to answer these questions.

Like so many of the lessons of history, the precedents established for aerial warfare during World War I were often conflicting and contradictory. Early in the war the British had learned the vital importance of producing aircraft of superior performance. In January 1916 the Royal Flying Corps established a "hard and fast rule" prohibiting reconnaissance missions without an escort of at least three other aircraft until the British developed

2. Director of Air Service to Chief of Staff, 25 April 1919, NA, DMA Admin. Box 4, 321.9 Air Service Program. See also Final Report of Asst. Chief, Air Service, AEF, 11 Nov. 1918, NA, WWI Orgn. Records, A.S. Hist. Records Box 1.

a type "as good or better than the German Fokker." [3] Here was vivid proof: one German aircraft of superior performance characteristics required a concentration of four inferior British aircraft. A superior weapon gave the Germans a definite advantage. The 1916 lesson of the Fokker was subsequently incorporated in official British military policy when a paper on aerial warfare drawn up by the General Staff emphasized the importance of flying in formation (i.e., quantity) except in the case of aircraft superior to those of the enemy (i.e., quality). [4] Thus, even before the United States entered the war, British policy implicitly recognized the importance of better weapons rather than more weapons. It is not to be supposed, however, that recognition of this principle implied universal adherence to it. Even among highly placed British officials there was a tendency to speak in terms of quantity rather than quality. No less a person than Gen. Sir David Henderson of the British Air Board advised Colonel Bolling to determine strength in pursuits on the basis of a three-to-one ratio. That is, he recommended sending three such aircraft into the field for every one sent by the enemy. General Henderson was certainly not unaware of the factor of quality, but his comment is representative of the tendency to think in terms of numbers rather than performance. [5]

The conflict of precedents which marked British experience was repeated in the United States. The decision to develop a new and superior engine rather than copy Allied designs was of itself a decision favoring quality over quantity. It is true that the Liberty engine was subsequently blamed for many of the failures of the program for aviation when it proved difficult to convert airframes to mount the new standardized engine. Nevertheless, there can be no doubt but that the initial decision to develop in the United States an engine far more powerful than any in use on the front was a decision which recognized the importance of better weapons rather than more weapons. Unfortunately for the success of aircraft production in the United States, the decision to build superior engines was apparently not made as a logical consequence of a conscious and widely accepted policy stressing quality over quantity.

3. RFC Headquarters Memo, 14 Jan. 1916, quoted in H. A. Jones, *The War in the Air* (5 vols., *2–6;* Oxford, Clarendon Press, 1922–37), *2,* 156.

4. British General Staff Memo, "Fighting in the Air," March 1917, quoted in full in Jones, *The War in the Air,* Vol. *3,* App. XI.

5. See above, Chapter 4. For a similar expression in the United States, see Memo, Lt. Col. V. E. Clark to Chief Signal Officer, 12 Sept. 1917, reprinted in full in *Aircraft Production,* Hearings before subcommittee of Senate Military Affairs Committee, 65 Cong. 2 Sess., 15 July 1918, *2,* 800.

Instead, it appears to have been an expedient or random act. That it did not result from conscious policy is suggested by the parallel decision to put the inferior and obsolescent DH-4 rather than the DH-9 into mass production. Here, side by side, were two violently conflicting policies. One favored better weapons, the other favored more weapons. Which precedent was the postwar Air Service to follow?

If the decisions of officials controlling production during World War I left no clear precedent, such was not true of the Air Service, AEF. Summarizing the lessons of aerial warfare learned by the air arm in France, the Chief of the Air Service, AEF, emphasized the critical importance of progressive development of design in maintaining a force of aircraft superior to that of the enemy on the front. An aircraft which dominated the air one day, he reported, might be "totally obsolete" six months later. Superiority in the air, he concluded, could be maintained only by constant initiative, encouraging inventions leading to new types, and, where necessary, abandoning unsuccessful models even after they have been brought into production.[6] Having been exposed to the cutting edge of the enemy's weapon, officers in the Air Service of the AEF felt that there was no question but that better weapons were of greater importance than more weapons. The vigor of this assertion was sharply underlined by the facts. When the war ended, the Allied forces in the theater of combat possessed almost twice as many aircraft as the enemy. Yet, despite this wide margin of numerical domination, officers in the AEF were reported to believe that the Germans "essentially overcame" their deficiency.[7] The Air Service, AEF, had learned the importance of superior weapons. Whether one measured that superiority in terms of more flying time for every airplane in action because of superior maintenance on the ground or in terms of superior performance in the air be-

6. "Final Report of the Chief of Air Service, AEF," *Air Service Information Circular, 2,* No. 180 (15 Feb. 1921), 37–8. See also draft version of text, NA, WWI Orgn. Records, A.S. Hist. Records Box 1.

7. Col. G. W. Mixter and Lt. H. H. Emmons, *United States Army Aircraft Production Facts* (Washington, GPO, 1919), pp. 57–8; and E. S. Gorrell, *The Measure of America's World War Aeronautical Effort* (Northfield, Vt., Norwich University, 1940), p. 58. The figures cited by these two sources disagree but not significantly. Jones, *The War in the Air,* Appendix Volume, App. XXXI, notes that the records in the Potsdam *Reicharchiv* show actual strength of the German air arm to have been but two-thirds of nominal strength officially indicated. This would appear to confirm the belief that it was to a great extent the superior performance of German aircraft rather than the greater number of hours they flew per airplane on the front which helped the Germans to make up for their over-all numerical inferiority.

cause of better design, the conclusion remained: better aircraft proved more desirable than more aircraft.

The combat-tested precedent of the air arm in France rather than the conflicting precedents of production on the home front shaped the course of the postwar Air Service. The chief of the new service reiterated the opinions of his predecessors in the AEF. Upon the organization for developing aircraft he placed "the larger portion of the burden of preparedness." Only by "continuous development," the chief reported, would it be possible to meet the enemy "on equal ground." [8] The phrases themselves reflect the extent to which experience in combat had centered attention upon the importance of progressive design in attaining future superiority. Perhaps the most telling postwar indication of the significance attached to superior weapons is to be found in the declared policy concerning the budget for the new peacetime air arm. Rather than procure and maintain a great number of aircraft for purposes of defense, the Air Service proposed to devote a large portion of its limited appropriations to experimental development and limited procurements for service testing. The policy of the Air Service would be, the chief reported, to "worry along" with obsolete equipment, spending every possible dollar from available funds for development and engineering rather than for large-scale orders for the production of aircraft from existing designs.[9]

Budgetary support represented compelling evidence that the postwar Air Service had learned from wartime experience the importance of weapons of superior performance. Was it equally true that the service learned the importance of a system for converting scientific and technological advances into weapons to ensure continued superiority?

The postwar Air Service, like the General Staff, failed to carry over from the war the concept of an agency attached to the highest staff for the purpose of exploiting science. To be sure, regulations and directives of the Air Service did make the Engineering Division officially responsible for keeping in touch with inventors and manufacturers, yet nowhere was there an organization specifically charged with the exploitation of scientific advances.[10] The Air

8. Report of Chief of Air Service, *Annual Reports of the War Department*, 1920, *1*, 1489.

9. *Ibid.*, pp. 1465, 1490.

10. "Outline of Organization and Functions of Office of Director of Air Service, U.S. Army," 5 Nov. 1919, and "Organization of Office, Chief of Air Service," 1 Dec.

Service, like the General Staff, lumped all functions of supply into a fourth section of the staff (similar to G-4) which was better equipped to handle problems of quantity than problems of quality.[11] Although the Science and Research Division disappeared with the peace, the National Research Council (NRC), from which it originally derived, remained. But the NRC suffered almost the same fate as its offshoot. When the Air Service failed to establish an organization at a high echelon to ensure continual and aggressive liaison, the NRC became a passive pool of talent available for calls which seldom came. Not until nearly 20 years later when another war was imminent did the NRC play an active role in assisting the air arm.

Insofar as it was concerned with problems of interest to the air arm, the National Advisory Committee for Aeronautics (NACA) fared considerably better in the postwar period than did either the Science and Research Division or the National Research Council. The survival of NACA was largely the result of its status, independent of the military air arm and blessed with separate appropriations. It was this very independence, however, which lessened the NACA's worth as an agency for scientific liaison for the postwar Air Service. Valuable as the services of NACA undoubtedly were, they did not answer the requirement for an agency on the staff of the air arm to bring into focus the advances in the broad field of the sciences for specific utilization in perfecting weapons.

In terms of agencies for positive action in calling attention to radically new findings in science, the air arm's heritage from World War I was slight indeed, which is somewhat surprising in view of the promising progress made during the war.[12] In terms of or-

1921, AAF Archive M1063-2, 3. See also "Functions of the Air Service," draft regulation (undated, ca. 1919), NA, BAP Hist. Box 7, 314.7 BAP History, General.

11. Even though the Air Service failed to establish an effective agency for liaison with the field of science, the importance of this function was not unrecognized by individuals within the air arm. In a study prepared as a rebuttal to a General Staff paper on the problem of military supply, an air officer (Maj. B. Q. Jones) analyzed the whole problem of dynamic design or the constant flux of aircraft types to maintain superiority. This special study treated the problem with such remarkable breadth that years later it remains an unusually clear statement of the present-day problem of aerial weapons. See "The System of Supply," A.S. Study, 5 May 1921, NA, DMA Admin. Box 20, 835 Executive.

12. The Science and Research Division was, on paper, merged with the Engineering Division. As a consequence the Engineering Division floundered around during the first few years directly following the war trying to determine its role with regard to fundamental research. By assigning the Science and Research Division to Engineering, Air Service authorities not only stripped the policy staff of its agency for scientific liaison but almost certainly ensured the subservience of science to engineering. For a discussion of the Engineering Division's postwar problems, see

ganizations for applied research and engineering, wartime experience contributed a more substantial heritage.

In the course of the war the Bureau of Aircraft Production built up a large organization for engineering. The Engineering Division, as it came to be called, ultimately employed more than 2,000 people.[13] In addition, during 1917 and 1918, the division had acquired an imposing array of facilities, including two airfields and several laboratories for experimental development and testing of engines, propellers, and airframes. Impressive though these facilities may have appeared in contrast with the makeshift establishment of early 1917, the Engineering Division which emerged from the war was an organization established primarily for production rather than for experimental development. Moreover, the models manufactured on a large scale were, in every instance, modifications of foreign designs. Of the more than 2,000 people in the division at the Armistice, approximately 75 per cent were entirely concerned with problems of production. Only a minute fraction of the engineering staff actually engaged in work on experimental development.[14] It was this small group of experimental engineers which was to become the nucleus of the postwar organization for applied research and development in the air arm. The cadre of engineers undoubtedly represented a valuable asset in meeting the problems of applied research in peacetime. Nevertheless, just as the organization for production during the war was ill suited for the role of experimental development in peace, so, too, the methods and procedures of the war were of little use to the Engineering Division in peacetime.

During the war the development of aircraft, no less than the production of aircraft, stemmed from European requirements. But following the Armistice, the Technical Section, AEF, disappeared, leaving no operational organization to provide information upon which to formulate design. The postwar Engineering Division, no longer able to call upon the AEF for this data, was forced to devise new practices and procedures for determining

"Rotary-Wing Aircraft in the Army Air Forces: A Study in Research and Development Policies," 1946, Wright Field Hist. Office.

13. The engineering organization of the air arm had a varied series of designations: The Airplane Engineering Department of the Signal Corps Equipment Division became the Airplane Engineering Division in January 1919 when the engineering organization merged with the Technical Section of DMA. In March 1919 the organization became the Engineering Division and kept that name for several years.

14. Photostat of questionnaire, Chief, Info. Group, to Engineering Div., 7 May 1919, completed by Engineering Div., Wright Field Hist. Office Files.

functional objectives. The Engineering Division was to devote the better part of 20 years following the Armistice to perfecting a system for converting these tactical and strategic objectives into detailed specifications.[15] It is somewhat ironic that the Technical Section, AEF, which had barely begun to evolve a system to formulate requirements, was allowed to dissolve, leaving behind scarcely a scrap of detailed information about its operating procedures.[16] For, as a consequence of the demise of the Technical Section, the Engineering Division in the United States was forced to acquire skills and practices which, at that same time, the Technical Section was abandoning. Because it did not have an effective system for ensuring retention of operational procedures evolved in war, the Engineering Division entered the era of peace ill equipped to handle the function thrust upon it.[17]

The postwar air arm lacked an effective organization for scientific liaison and inherited a partially equipped organization for applied research. Despite some substantial facilities for testing and development as well as a nucleus of personnel for engineering, the air arm faced the peace as it had entered the war—without an operational organization to determine the direction engineering development should take. In the final analysis, doctrine, or the accepted concept of the mission to be performed by the aerial weapon, would inevitably determine the direction of development. To understand in full the heritage of World War I it will be necessary to consider air doctrine.

15. For an illustrative study of the problems encountered in formulating functional objectives—or military characteristics, as they are called—and the problems encountered in reducing these objectives to specifications for types of aircraft, see "The System of Supply," A.S. Study, 5 May 1921, NA, DMA Admin. Box 20, 835 Executive. See above, p. 154 n. 11.

16. Even the "Air Service, AEF, History," NA, WWI Orgn. Records, GHQ, AEF File, which was specifically designed to record the lessons of the war, contains only the sketchiest account of operating procedures in the Technical Section.

17. A careful search in the files of the Engineering Division and the Air Service Central Files covering the formative period immediately following the Armistice revealed almost no evidence of extensive reliance on wartime studies of organization when the new Air Service was established.

## Chapter X. The Postwar Air Service and Air Doctrine

THE BODY OF experience from which the postwar air arm had to draw was extremely small. Aerial operations in the AEF were confined to six or eight months of activity. Thus, when a board of officials met shortly after the Armistice to consider the question of organization only one officer who had actually commanded an aerial unit in combat could be found in the whole United States. It is interesting to note that this officer was Maj. Carl Spaatz who became the Commanding General of the Army Air Forces after World War II.[1] Experience in terms of personnel was, at best, elusive. Even as more officers with service in combat became available, their utility in assisting in the formulation of doctrine was sharply circumscribed by the degree to which they were articulate about their experience. Even the most thoughtful and expressive were limited by the operations of the Air Service, AEF, which were confined almost entirely to service in a role of army-cooperation or close-support.

At the Armistice the Air Service, AEF, had 18 squadrons for observation but only one squadron for night bombing assigned on the front. Even the lone squadron of night-bombers belied its name: it was not equipped with twin-engine, long-range Handley-Page airplanes but consisted of 18 obsolete DH-4 aircraft for observation and an equal number of British FE-2 aircraft, a type which had reached the peak of its production before the DH-4. The squadron was scarcely equipped for strategic bombardment.[2] Perhaps the most significant fact about the lone squadron of night-

1. Memo, Asst. Chief of Training, DMA, to Board of Organization, 18 Dec. 1918, AFCF, 334.7 Board on Organization of the Air Service in Peace.

2. The average bomb load of a Handley-Page amounted to approximately 1,325 lbs. The average DH-4 bomb load was 232 lbs. Thus, even disregarding the wide difference in range, it is evident that the DH-4 was little suited to strategic bombardment. See "Statistical Analysis of Aerial Bombardment," General Staff Statistics Branch, Report No. 110, 7 Nov. 1918, National War College Library, U226, U60, 1918. See also E. S. Gorrell, *The Measure of America's World War Aeronautical Effort* (Northfield, Vt., Norwich University, 1940), p. 29.

bombers was not so much the inadequacy of its equipment as the brevity of its experience. Assigned to the front on 9 November 1918, the squadron was in active service only two days before the Armistice. Moreover, this single squadron of night-bombers was officially demobilized 5 December 1918, less than a month after it entered combat.[3] When the war was over there was no familiar body of operational experience with strategic bombers on which to draw in formulating doctrine for the postwar establishment.

Officers in the Air Service, AEF, were primarily concerned with aviation for army-cooperation. Col. T. D. Milling, who had led the air units of the First Army in France, held opinions typical of the majority of airmen in the higher echelons of the AEF. In reporting his conception of a proper postwar air arm Colonel Milling emphasized one point: "The Air Service is of value to the military establishment only insofar as it is correlated to the other arms."[4] This was the opinion of an important leader in the air arm. The opinion was not unique. Col. F. P. Lahm, who led the aerial units of the Second Army in France, felt that the Air Service, AEF, had not had sufficient operational experience to permit one to reach any definite conclusions regarding doctrine for the future. Nonetheless, having judiciously recognized the narrow base of operations in the AEF, the colonel proceeded to declare that "the main function of aviation is observation and all hinges on that program."[5] This contention, hazarded in the face of the colonel's avowed denial of its validity, almost certainly reflects the limited range of his experience. More than half of the aviation units with the Second Army were equipped with the obsolete DH-4 while the remainder were equipped with pursuits of French and British origin.[6]

When the Chief of the Air Service, AEF, came to prepare his final report, it was perhaps natural for him to perpetuate the opinions of some of the officers who had commanded the operational units of the United States' forces in France. In any event, whether compiled on an objective basis or not, his final report repeated the opinions of the principal air officers of the First and Second armies. The war "clearly demonstrated," the Chief of the Air

3. Gorrell, *The Measure of America's World War Aeronautical Effort,* p. 33.
4. Copy, memo, Col. T. D. Milling to Chief, Air Service, AEF, 9 Jan. 1919, in "Lessons Learned," "Air Service, AEF, History," Series A, Vol. *15, ca.* 1919, NA, WWI Orgn. Records, GHQ, AEF Files.
5. Col. F. P. Lahm to Col. E. S. Gorrell, 7 May 1919, NA, WWI Orgn. Records, A.S. Hist. Records Box 572, 452.1.
6. Gorrell, *The Measure of America's World War Aeronautical Effort,* p. 43.

Service believed, that the function of observation was "the most important and far reaching" which an air arm operating with an army would be called upon to perform.[7] This report, which appeared after the Armistice, is strangely reminiscent of the opinions held by the Chief Signal Officer in 1915. At that time, it may be recalled, he had declared that observation was "the useful, approved, and most important work" of the airplane.[8] Here, then, was the completed cycle.

The United States entered the war without a clearly defined doctrine of aerial warfare. Insofar as a doctrine had been formulated, it favored a role of army-cooperation or close-support rather than an offensive or strategic role. Because of this policy, production during the war favored aircraft for observation rather than night-bombers. As a consequence, the experience of the war was with observation rather than bombardment. In deliberating on this experience, postwar planners found themselves precisely where they started. If the air weapon was to develop beyond this limitation, some way had to be found to escape the self-perpetuating character of this cycle.

British experience offered a substantial precedent showing that it was perfectly possible to escape the merry-go-round of doctrine influencing production and determining experience in operations which formed the basis of subsequent doctrine. The British projected doctrine beyond experience into the realm of possibilities. The Royal Flying Corps as planned in 1916 was to consist of 86 squadrons. Of these only 10 squadrons were to be of long-range bombers. By 1917 the air force planned 179 squadrons, of which 66 were to be used for strategic bombing. By the summer of 1918 planned distribution of British strength allocated 32 per cent of the available aircraft to the Independent Force for strategic operations.[9] Even though the limitations of aircraft production never permitted a large-scale proof of the concept of strategic bombardment, British officials were able to secure approval and support for their doctrine from the highest authorities, both military and political.

In the United States efforts at formulating an authoritative

7. "Final Report of the Chief of Air Service, AEF," *Air Service Information Circular*, 2, No. 180 (15 Feb. 1921), 49.

8. *The Service of Information*, Circular No. 8, Office, Chief Signal Officer, 1915, p. 23.

9. W. A. Raleigh, *The War in the Air* (Oxford, Clarendon Press, 1922), *1*, 452–3, and H. A. Jones, *The War in the Air* (5 vols., *2–6;* Oxford, Clarendon Press, 1922–37), *6*, 171–2.

doctrine met with less success. Since air doctrine determines the direction of development in aircraft one might logically assume that military officials would devote much effort to the formulation of doctrine at the end of the war, but this was not the case. Neither the Commander-in-chief of the AEF nor the Chief of the Air Service, AEF, pointed out any comprehensive lessons of aerial warfare in their final reports.

To judge from General Pershing's report the air weapon was utterly insignificant. Apart from an occasional sentence of the aircraft-also-flew variety in accounts of operations by the ground forces, Pershing's report failed to speculate on the mission of aviation and neglected to appraise the role of aerial warfare. It confined all mention of the air arm to a single paragraph commending the aviators for their courageous deeds and daring.[10] The report of the theater commander was most general and quite above detailed consideration of technicalities, an omission which may have been unavoidable in the postwar rush to complete the report. But in view of the revolutionary importance of the weapons slighted, the neglect is somewhat disconcerting.

The final report of the Chief of the Air Service, AEF, was even more surprising. Although this document did not appear in published form until February 1921, it was originally submitted in March 1919. Again the urgency of events surrounding the demobilization of the AEF undoubtedly influenced the character of the paper. Far from representing a thoughtful and carefully reasoned study of the role of the air weapon in the war, this report was a jumbled narrative account emphasizing the difficulties encountered without critical appraisal or thorough analysis. More significantly, the report considered all of the various functions of the air arm individually, drawing occasional conclusions regarding supply, operations, or training but never assembling these conclusions into one comprehensive whole. Nowhere were the lessons of the war presented in concerted form for the instruction of the future. Moreover, despite the novel character of the air weapon, the report showed no systematic effort to derive appropriate doctrines for aerial warfare. There is some evidence to indicate that the report did not represent even so much as an exhaustive compila-

10. *Final Report of Gen. John J. Pershing, Commander-in-chief, American Expeditionary Force* (Washington, 1920), p. 76. It is interesting to note that this report, which mentions aviation in a section entitled "Supply, Coordination, Munitions, and Administration," devotes about the same amount of space to religious activities as it does to aviation and tanks combined.

tion of opinions held by officers in the top echelons of the Air Service.[11]

The critical importance of formulating doctrine for the air weapon was not everywhere so neglected as it was by the air arm itself. Officers on the General Staff in Washington were deeply interested in the problem of deriving lessons from the experience of combat to guide the conduct of military operations in the future. To this end the General Staff undertook to prepare a series of statistical reports for publication as source materials upon which to base studies on policy. Since these reports were consciously intended for use in formulating doctrine, they could scarcely fail to influence postwar military thinking. For this reason they merit careful attention.

Typical of the studies drawn up by the General Staff to evaluate the air weapon was one entitled "Statistical Analysis of Aerial Bombardment" which appeared only a few days before the Armistice. By its own admission the report was based on fragmentary data, "occasional" photographs, accounts by prisoners of war, and information "too meager and too unreliable to serve as a basis for any quantitative analysis." [12] Having made this admission, the compilers of the report apparently believed they were free to draw entirely valid conclusions from the fragmentary evidence available. From limited facts the analysts of the General Staff drew some rather startling deductions. The material destruction to be expected from bombardment "must necessarily be small," they felt, because of the limited weight of bombs carried and the scant number of bombers which could be maintained on the front. Using this frame of reference, the report continued, "From the standpoint of destruction, the function of the bombing plane is, in the last analysis, practically the same as that of long-range artillery." By utterly ignoring the factor of range, the statisticians of the General Staff virtually reduced the bomber to a field gun. Then, to prove the point, they cited figures to show that 300 British aircraft dropped a daily average weight of explosive about equal to the explosives thrown by the total number of 155 mm. guns availa-

11. Some of the working papers used in preparing the final report of the Air Service, AEF, are available in the National Archives, WWI Orgn. Records, A.S. Hist. Records Box 1 ff. The omission of Brig. Gen. William Mitchell's point of view is noteworthy.

12. "Statistical Analysis of Aerial Bombardment," General Staff Statistics Branch, Report No. 110, p. 21, 7 Nov. 1918, National War College Library, U226, U60, 1918.

ble in the same period. "In other words," the study concluded, it took "two squadrons of bombing planes to equal the work of one 155 mm. gun." [13] If it were at all necessary to demolish this patently fatuous argument, the report itself contained all the essential facts. The statistics upon which these surprising deductions were based had been selected from British bombing operations before the formation of the Independent Force. Moreover, in the missions analyzed, only five per cent of the explosives dropped were heavy demolition bombs. The remainder were almost exclusively fragmentation bombs used primarily for antipersonnel work. Similarly, of these same missions, only five per cent were against industrial targets; most of the raids were made on such objectives as troop billets, communications lines, and ammunition dumps.[14]

The study by the General Staff recognized that the targets attacked by British bombers had been increasingly industrial in the period following that from which the statistical analysis was made. But the analysts deprecated this trend by pointing out that the total area of damage was slight. If the craters of all the bombs dropped were to be concentrated in one place, they noted, the total area covered would amount to no more than some 500 square feet with an additional area of damage resulting from blast.[15] Such statistical perversions suggest a strong predisposition in the General Staff to ignore the strategic potential of aviation.

The report on bombardment was by no means an isolated instance of the statistical techniques used by the General Staff in meeting the problem of air doctrine. Another item in the series of studies was entitled "Distribution of Service Planes As Related to Military Policies." [16] This report represented a conscientious effort to prepare the necessary groundwork before undertaking to formulate a doctrine on aerial warfare for the United States. As a systematic approach to the problem of doctrine it was probably unique. Variations in the composition of the air forces on the front, the report contended, made it obvious that there was a lack of common doctrine on the function of military aviation. In support of this position the study presented the following chart showing the composition of the major air forces in August 1918.[17]

13. *Ibid.,* p. 23.
14. *Ibid.,* pp. 7, 14.
15. *Ibid.,* p. 23.
16. Photostat copy, "Distribution of Service Planes As Related to Military Policies," General Staff Statistics Branch, Report No. 2, undated, *ca.* Aug.-Sept. 1918, NA, WWI Orgn. Records, A.S. Hist. Records Box 512, 452.1.
17. *Ibid.,* Diagram XI, shown only in part.

PERCENTAGE OF AIRCRAFT BY FUNCTIONAL TYPES ON
FRONT, AUGUST 1918

|  | Pursuit | Observation | Bombardment |
|---|---|---|---|
|  | % | % | % |
| British | 55 | 23 | 22 |
| United States' | 46.5 | 46.5 | 7 |
| Italian | 46 | 45 | 9 |
| German | 42 | 50 | 8 |
| French | 34 | 51 | 15 |

From this evidence the statisticians of the General Staff concluded that the several forces represented "widely divergent theories" on doctrine. The existence of these divergent theories made it imperative, they concluded, that careful studies be made to establish "a basis for intelligent judgment" in formulating doctrine. On the basis of the figures presented in the above table, the statisticians decided that the two divergent theories of doctrine were best typified by the British on the one hand and the French and Germans on the other. The British theory held that aviation was a combatant arm whose mission it was to seek out and destroy the enemy. The French and German theories held that military aviation was primarily concerned with reconnaissance and adjustment of artillery-fire.

In the absence of any prior systematic attack on the problem of aerial doctrine this study by the General Staff was a commendable effort. Nevertheless, the statistics presented in the report were intrinsically faulty; any lessons derived could not be other than erroneous. The statisticians freely admitted that limitations of production in some instances may have influenced the character or composition of the air forces actually assigned to the front. Then, having made this admission, they went right ahead and drew conclusions regarding the "divergent theories" which the figures reflected.

Officers in the Air Service were quick to point out the faulty premise of the analysis.[18] The strength of units at the front was not a true index of the air doctrines of the several Powers. A more accurate index, the airmen suggested, would be found in the strength of the units planned by the combatants. The following table, showing the strength in squadrons which the Allies hoped

18. *Ibid.* See enclosures with photostat copy of report, for example, Lt. Col. Ralph Royce to Col. E. S. Gorrell, 6 May 1919, "a cleverly arranged paper which leads to false conclusions," and Lt. Col. E. L. Naiden to Gorrell, 7 May 1919, as well as Gorrell's own vigorous marginal notation, "A rotten paper."

to have along the front by July 1919, indicates how markedly different actual strength was from planned strength.[19]

PROGRAMS SUBMITTED BY THE ALLIES SHOWING
DISTRIBUTION OF SQUADRONS BY FUNCTIONAL
TYPES PLANNED FOR JULY 1919

|  | *Pursuit* | *Observation* | *Bombardment* |
|---|---|---|---|
|  | % | % | % |
| British | 36 | 10 | 54 |
| United States' | 30 | 50 | 20 |
| Italian | 20 | 49 | 31 |
| French | 21 | 46 | 33 |

By comparing the above table with the preceding one one can easily see that the conclusions reached in the study by the General Staff as to the divergent theories of doctrine no longer appear so valid. After making due allowance for the circumstance that the strength of observation units was directly proportional to the number of troops in the field and therefore an arbitrary commitment, the "divergent theories" are more difficult to perceive. More significantly, in every instance the ratio of pursuits to bombers in the planned units showed a marked increase in favor of bombardment. Far from showing divergent theories on doctrine, the planned distribution of units reveals a remarkably common trend, with the emphasis shifting from strength in pursuit to strength in bombardment. Strength in observation units computed on a troop basis remained somewhat more constant. But these computations, while more revealing than actual strength on the front, were no certain index of Allied doctrine of air power. As indicated in an earlier chapter, even the final program planned for the Air Service, AEF, was based on the capabilities of production rather than on doctrinal concepts.[20] What was true of the AEF may well have been true of the Allies. The British Air Ministry consciously scrapped long-range plans for immediate advantage in October 1918 when the German collapse became imminent.[21] Similar factors may well have influenced all other Allied programs.

The evidence seems to show that the statisticians of the General Staff were repeating the pattern of their report analyzing bombardment by selecting facts from which to derive predetermined conclusions. While the report actually made no overt conclusions

19. Percentages calculated from Lassiter Board Report, App. II, 17 March 1923, copy in Library of Industrial College of the Armed Forces.
20. See above, Chapter 9.
21. Jones, *The War in the Air, 6,* 172–3.

as to a doctrine of air power, it did, nevertheless, establish a misleading premise. From this premise faulty doctrine might easily be derived by subsequent analysts uncritical of the assumptions employed in compiling the original study. Officers in the Air Service recognized the erroneous character of the statistics used by the General Staff, but having done so they failed to take any action which would draw conclusions and formulate doctrine on a revised premise.[22]

Interest on the part of the General Staff in finding an objective basis for determining the most advantageous role of the air weapon was not confined to a single study or even to a single technique. Another attack on the problem sought to find some meaning in the relationship represented by the total number of aircraft available on the front per hundred thousand rifles in combat. This report appeared during October 1918 under the caption "Ratios of Aerial Equipment to Army Strength." The Chief of the Air Service, AEF, hurriedly contradicted it on the same grounds used in demolishing the report described previously. The statisticians of the General Staff had assumed that the aerial strength of the Allies on the front in the spring and summer of 1918 represented an accurate measure of their policies based on operational experience. In reality, the Chief of the Air Service contended, nothing could be farther from the truth; Allied strength in airplanes was based on availability rather than on desired ideals.[23]

After belaboring the erroneous conception regarding Allied policy utilized by the report of the General Staff, the Chief of the Air Service submitted a report in reply.[24] His counterblast called specific attention to the futility of searching for significance in the relationship between strength in the air and the number of

---

22. Despite the caustic remarks of officers from the air arm regarding attempts by the General Staff at formulating doctrine from a statistical premise (see above, p. 163 n. 18), the files of the period contain no evidence to show that the criticism was creative as well as caustic. There is, however, other evidence of faulty techniques used by the General Staff Statistics Branch in studies of the air arm. See, for example, Asst. Chief of Training to Director, Air Service, 13 March 1919, AFCF, 059 Statistics, Miscellaneous, in which analysts of the Statistics Branch are charged with an "attitude antagonistic to the Air Service."

23. Maj. Gen. M. M. Patrick, Chief, Air Service, AEF, to J. D. Ryan, Second Asst. Secretary of War, 5 Nov. 1918, letter drafted by Col. E. S. Gorrell, NA, BAP Exec. Box 6, 026.4 BAP. A similar comparison is mentioned in A. Sweetser, *The American Air Service* (New York, D. Appleton, 1919), pp. 234–5, showing that the weight of strength in aircraft compared to strength in rifles was a common evaluation at the time.

24. Maj. Gen. M. M. Patrick to J. D. Ryan, 5 Nov. 1918, enclosure labeled "Calculation made 3 Nov. 1918," NA, BAP Exec. Box 6, 026.4 BAP.

rifles on the front. Strength in rifles, or the troop basis, he asserted, was not an entirely proper reference from which to calculate the composition of an air force. Instead he proposed three criteria: units for aerial observation in proportion to the strength of units on the ground; pursuit in proportion to the enemy strength; and bombardment on still another premise:

> The amount of bombardment aviation desired depends neither upon the size of our own ground army nor upon the enemy's air service, but solely upon a nation's ability to place this excess, so-called Air Service luxury, in the field after having provided the two essentials of observation aviation with an adequate amount of pursuit aviation.[25]

The statement on the composition of the air arm is particularly noteworthy in that it repeats much of the same language used in the policy-proposing report of the Bolling Mission in the summer of 1917, which in itself was more or less a paraphrase of the ideas expressed earlier by Sir David Henderson of the British Air Board.[26] Like the earlier opinions of Bolling and Henderson, the report of the Chief of the Air Service represented a relatively clear statement of policy. He considered that aircraft for pursuit and for observation were "essential" but that the size of the army and the activity of the enemy determined and limited the number required. He felt that there was no ceiling on bombardment aircraft save capacity to produce. Here, inferentially, was a statement of air doctrine. Squadrons of bombers should go on increasing in numbers limited only by the nation's facilities and resources. The Chief of the Air Service appeared to imply agreement with Lt. Col. V. E. Clark, who had contended that systematic bombardment would end the war sooner than sending one or two million men to line the trenches.

If the counterblast to the General Staff which the Chief of the Air Service, AEF, sent to the United States implied an aerial doctrine favoring unlimited bombardment, did the Air Service fully recognize the significance of this implication? Did the report of the Air Service connote a conscious derivation of doctrine, or did it merely seize upon the convenient language in the opinions of Henderson and Bolling without entirely appreciating their doctrinal allusions? [27] All the available evidence appears to favor the

25. Report of 3 Nov. 1918, NA, BAP Exec. Box 6, 026.4 BAP.
26. See above, Chapter 4.
27. The historical problem raised here is made doubly interesting because the same man, Col. E. S. Gorrell, who drafted the report of 3 Nov. 1918 for the Chief,

latter view.[28] If occasional officers in the Air Service did favor a policy emphasizing unlimited bombardment, they seldom formulated their views in terms of a comprehensive doctrine of air power. If the Chief of the Air Service believed in strategic bombardment, as one might infer from his rebuttal to the report of the General Staff, the Air Service failed to reduce his views to official doctrine when subsequently publishing the "lessons learned" from the war.

During the first few years immediately following the war the Air Service issued a number of studies on aerial warfare in an effort to profit from the recently completed operations in France. One of these studies, "Notes on the Characteristics, Limitations and Employment of the Air Service," sought to define some of the broad operating principles of an air arm. The functions of the aerial weapon, according to this publication, should include aid to the Infantry, fire-control for the Artillery, reconnaissance for the staff, and destruction of the enemy's air arm. Generally speaking, the study concluded, airplanes should assist in "deciding actions on the ground" while preventing the enemy from doing the same.[29] This publication was characterized by a limited range of thinking which closely followed the limited scope of operations conducted by the Air Service, AEF, in contrast to the far more extensive experience of other Powers. The boundaries of experience acquired by the Air Service were never so clearly defined as they were when expressed in the semiofficial postwar reflections on aerial warfare: "Whatever the future development of aviation may be, up to the end of the war in 1918, its most important functions had proved

Air Service, AEF, was a member of the Bolling Mission and may have written or shared in writing the Bolling report of August 1917. Gorrell may have harked back to the Bolling report for words to frame his reply, but his subsequent actions (see text below) indicate that his concept of doctrine must have changed during the year from the fall of 1917 to the fall of 1918.

28. Some officers in the Air Service may have been ardent advocates for strategic bombardment at the end of World War I, but if they were they left little written evidence of their point of view. Even Mitchell, leading proponent of the bomber in the public eye, in 1920 advocated offensive aviation in the proportion of 60% pursuit, 20% attack, and 20% bombardment. See Asst. Chief, Air Service, to Chief, Air Service, 16 Sept. 1920, NA, Air Service Finance Advisory Board Box 10, Case No. 179. For an example of the objective and scholarly analysis which other air arms devoted to the problem of doctrine, see Lt. Col. A. P. Voisin, *La Doctrine de l'aviation française de combat au cours de la guerre* (Paris, Berger-Levrault, 1932).

29. "Notes on the Characteristics, Limitations and Employment of the Air Service," *Air Service Information Circular*, Vol. 1, No. 72 (12 June 1920). Although carefully labeled as "unofficial" because it lacked approval by the General Staff, this study and others in the series were widely distributed throughout military circles with consequent influence on concepts of air doctrine.

to be securing and transmitting information concerning developments in and beyond the line of battle." [30]

By entering the war with a predisposition to see the role of the air weapon as primarily one of observation, the air arm had circumscribed the limits of its operational experience. Then, at the end of the war, the new Air Service in its postwar study pointed to this experience as confirmation and proof of the primary importance of observation. At the same time, even while cautiously suggesting that the future might reveal other roles for the air arm, the study made no mention of the function of strategic bombardment even as an embryonic development. Would air arm officers ever get off the merry-go-round?

The postwar publication of the Air Service on aerial doctrine was scarcely an objective approach to the problem. By the editor's own admission the principles or policies presented had been evolved from a report entitled "Notes on the Employment of the Air Service from the General Staff Viewpoint" and submitted for criticism "to those most expert" in the AEF.[31] If, as officers of the air arm had repeatedly pointed out, analyses made by the General Staff had shown a predisposition to ignore the strategic potential of bombardment, then surely the point of view of the General Staff was a faulty one from which to derive lessons for the future air arm. As for "those most expert" who criticized the study, did they represent a cross section of experienced air officers or only those who agreed with the doctrine presented?

The available evidence shows that during the postwar period the efforts of the Air Service to formulate doctrine were neither systematic nor objective. With the best of intentions the Air Service began a number of historical studies of wartime operations to "furnish valuable sources" from which to derive lessons on strategy, tactics, supply, training, and the like.[32] Unfortunately, few of these studies reached the stage of publication.[33] And, of equal significance, expressions of doctrine formulated in the few studies actually completed found no place in the official publication of the postwar Air Service when they were at variance with doctrine as-

30. *Ibid.*

31. *Ibid.* The original study, "Notes on the Employment of the Air Service from the General Staff Viewpoint," appeared in Feb. 1919. A photostat copy is available in AAF Archive, 1411–8. This copy is undated, and no authorship is indicated.

32. Report of Chief of Air Service, *Annual Reports of the War Department,* 1920, *1,* 1476.

33. For a discussion of the fate suffered by historical studies of World War I, see bibliographical note.

signing primary importance to observation. Forthright statements
in favor of strategic bombardment as made by such officers as
Lt. Col. V. E. Clark remained buried in the files. "There should be
no limit to the number of night-bombing machines supplied," Clark
wrote. "The policy should be to bomb the enemy at every vital
point until he is exhausted." The doctrinal position of this contention was clear enough, but it was ignored.[34] Had the studies
of the Air Service been systematic they would have mentioned this
position if only to controvert it. But for want of an objective and
authoritative method of formulating doctrine on air power, the
manifestly inadequate doctrine emphasizing the role of observation reached publication and consequent circulation while opposing points of view did not.

The need for systematic formulation of doctrine may well have
been recognized, but like so many functions of the air arm in the
postwar era, it was neglected for want of competent personnel to
conduct the necessary analysis. There were 431 officers at the important Paris headquarters of the Air Service, AEF, at the end
of the war. Of these, only one was an officer of the Regular Army.[35]
The inevitable result of this disparity appeared less than a year
after the Armistice. In the whole Air Service by July 1919 there
were only 234 regular officers, barely enough to man three full
squadrons if so assigned.[36] In any event, whether the inadequacy
of the doctrine of air power formulated after the war resulted from
insufficient personnel or from a failure to appreciate the vital importance of doctrine, the effects were the same. The Air Service did
not establish a system to ensure an objective analysis of wartime
operations, foreign as well as domestic, enemy and Allied as well
as those of the AEF. The product of this failure followed with
inexorable logic: the doctrine of aerial warfare which the Air
Service promulgated after the war fell ominously short of the
airplane's full potential.

Because the Air Service neglected to take the initiative in exploiting the full scope of the aerial weapon, doctrine in 1919 remained only slightly advanced over what it had been in 1917. If
the Air Service itself formulated a doctrine which utilized far less
than the maximum potential of aviation, military opinion out-

34. Lt. Col. V. E. Clark, "History of the Development of Airplanes During the
War," 1 Jan. 1919, NA, BAP Hist. Box 19, 452.1 Airplanes, General.
35. Final Report of Asst. Chief, Air Service, AEF, 11 Nov. 1918, NA, WWI Orgn.
Records, A.S. Hist. Records Box 1.
36. AAF Historical Study No. 39, "Legislation Relating to the Air Corps Personnel and Training Programs, 1907–1939," Dec. 1945, AAF Archive.

side the air arm could scarcely be expected to favor more advanced thinking.

The Dickman Board, a postwar group ordered by General Pershing to consider the lessons of the war, arrived at substantially the same concept of aerial warfare as that held by the Air Service: ground forces would continue to play the major part in both offensive and defensive operations, using the air arm as an "auxiliary." [37] The Dickman report set a pattern which was followed repeatedly. In 1919 Assistant Secretary of War Benedict Crowell headed a postwar mission to Europe to study the lessons of aviation learned by the Allied Powers. On his return he wrote: ". . . the primary purpose of war flying is observation." [38] General P. C. March, the Chief of Staff, was in substantial agreement. Despite his momentary wartime enthusiasm for strategic bombardment, after the war General March appeared more conservative. In writing of tanks, aircraft, and other innovations, he asserted flatly that their true value lay in their service as auxiliaries to the Infantry. "Nothing in this war has changed the fact that it is now, as always heretofore, the Infantry with rifle and bayonet that, in the final analysis, must bear the brunt of the assault and carry it on to victory." [39]

Newton D. Baker, the Secretary of War, was likewise inclined to follow the lead of the Air Service. He appeared to be supporting a policy of full exploitation for the air weapon by readily conceding that it would be "dangerous to attempt any limitation on the future based on the most favorable view of present attainments." [40] In spite of this seeming breadth of vision, the secretary consistently deprecated the role of bombardment while emphasizing the primary importance of observation. During the war, he said, the direct damage inflicted by bombs was "relatively small" and had "no appreciable effect" upon the war-making power of the enemy. Moreover, if bombing were outlawed, as the secretary thought "it plainly should be upon the most elemental ethical and

37. AAF Historical Study No. 25, "Organization of Military Aeronautics; 1907–1935," pp. 37–8, 1944, AAF Archive.

38. B. Crowell, *America's Munitions; 1917–1918* (Washington, GPO, 1919), p. 254.

39. Report of Chief of Staff, *Annual Reports of the War Department,* 1919, *1,* 473–4. Anyone familiar with the problems of critical shortages in materials and manpower in the economic mobilizations of both world wars can appreciate the relatively limited range of March's thinking when he capped his discussion of weapons with the declaration, "The war has shown that this country can, in an emergency, be self-sustaining in all respects for an indefinite period" (p. 475).

40. Report of Secretary of War, *Annual Reports of the War Department,* 1919, *1,* 68–71.

humanitarian grounds," then the value of aviation for observation would far outweigh the effectiveness of aviation for direct attack.[41] The confusion in Secretary Baker's thinking about the role of air power was not confined to an isolated instance. He described an aerial torpedo or guided missile developed by the well-known engineer Charles F. Kettering as "one of the most destructive weapons invented during the war." Furthermore, he asserted, perhaps somewhat prematurely, that this guided missile was already "a practical reality" being seriously considered as a replacement for heavy guns.[42] But then, like the spokesmen for the Air Service, the secretary failed to urge any positive course to develop this "most destructive" weapon or enlarge the latent scope of the air weapon in general, even after tacitly recognizing its significance. The secretary apparently ignored the implication of the facts he himself presented. In his eagerness to support the cause of peace, he seemed to be neglecting his role as Secretary of War.

With the Air Service, the Secretary of War, the Assistant Secretary of War, and the Chief of Staff all dwelling on the auxiliary role of aviation and stressing observation to be of first importance, it is not surprising that postwar training followed suit. Perhaps the best index of doctrine after the war is to be found in the teachings of the service schools rather than in the official pronouncements of doctrine made by high-ranking officials. What the next generation of officers was to think of aerial doctrine was probably more significant than what any given official thought on the eve of retirement, no matter how influential his role during the war may have been.

A textbook on the Air Service prepared for the Command and General Staff School in 1920 introduced the air weapon in the accepted frame of reference. "Teamwork with the ground troops," said the text, was "the basic idea" underlying the organization of aerial units.[43] The text presented the airplane as the servant of the ground arms. The relative importance of observation, said the author, himself an airman, was shown during the war. Germany,

---

41. *Ibid.* As chairman of the Baker Board on aviation in the mid-thirties, Baker persisted in deprecating the full potential of strategic bombardment even though the board recommended formation of the GHQ Air Force for "independent" bombing missions. See *Final Report of War Department Special Committee on the Army Air Corps,* 18 July 1934.

42. In a speech at Fort Worth favoring the League of Nations, Secretary Baker mentioned the aerial torpedo in some detail. See *Army-Navy Journal* (Washington, D.C., 24 May 1919), p. 1315. See also n. 41, above.

43. E. L. Naiden, *Air Service* (Fort Leavenworth, Kans., The General Service Schools Press, 1920), p. 6.

France, Italy, and the United States found "more observation squadrons were necessary for a well balanced army than any of the other types." [44] The text of the staff school conceived strategic aviation within the framework of the ground armies. Selection of targets was assigned to the army staff (G-3). The strategic force was to be formed of units left over *after* the tactical needs of armies and corps had been satisfied.[45] As a later staff school text phrased it, "Tactical bombing is . . . a necessity, while strategical bombing is . . . a luxury." [46]

Another index of official thinking is found in the types of aircraft procured following the war. The allocation of funds, always less than enough, provides a realistic picture of attitudes in the War Department toward air doctrine. The table below, showing numbers and types of aircraft annually accepted from manufacturers, reflects the shifting emphasis.[47]

AIRCRAFT ACCEPTANCES

|      | Pursuit | Bombers | Observation |
|------|---------|---------|-------------|
| 1920 | 112     | 20      | 1,000       |
| 1921 | 200     | 85      | 270         |
| 1922 | 60      | 25      | 200         |
| 1923 | 0       | 0       | 180         |
| 1924 | 35      | 0       | 127         |
| 1925 | 18      | 2       | 126         |
| 1926 | 83      | 1       | 176         |
| 1927 | 49      | 9       | 56          |
| 1928 | 69      | 34      | 151         |
| 1929 | 78      | 22      | 185         |
| 1930 | 108     | 30      | 122         |

44. *Ibid.*, p. 60. Naiden's attempt to show the importance of observation by reference to the wartime Powers is exceptionally curious. The author was one of the air officers who had protested vigorously at the attempts of analysts of the General Staff to show the importance of aviation for observation by reference to actual strengths on the front as an index of policy. See above, p. 163 n. 18.

45. Naiden, *Air Service*, pp. 69–70. While the text of this study boldly introduced the subject of aerial warfare by stressing the importance of observation and the servant-of-the-army concept, buried way back on p. 108 one finds in small print, so to speak, the following priority of bombardment targets: first, industrial objectives; second, railways; third, troop concentrations in the field. Could it be that the author was purposefully slipping in this strategic doctrine or were the implications of this obvious contradiction and inconsistency unrecognized?

46. *Corps and Army Air Service* (Fort Leavenworth, Kans., The General Service Schools Press, 1922), p. 26. This text, prepared for the staff school, bears the initials ELN, probably E. L. Naiden who wrote the text cited above. Subsequent texts for staff schools during the next 10 years taught substantially the same lessons.

47. Tabulation by General Staff Statistics Branch, 1 Nov. 1938, quoted in AAF Historical Study No. 44, "Evolution of the Liaison-type Airplane; 1917–1944," chap. ii, May, 1946, AAF Archive.

The character of instruction in staff schools and of procurement of aircraft in the postwar era clearly shows that the concept of aerial warfare carried into the peace was well below the maximum potential of the new weapon.[48] Whether by neglect or intention, the Air Service itself had determined the fundamental make-up of the heritage from the war years. Once having been promulgated and written into the record of the war, air doctrine proved hard to modify.

As early as 1921 the new Chief of the Air Service began to plead the cause of fuller exploitation of air power. Influenced, no doubt, by the writings of General Douhet, the Italian theorist, the Chief of the Air Service divided military aviation into two branches. On the one hand was air *service* or aviation serving in cooperation with armies, corps, and divisions. On the other was air *force*, which embraced all aviation of an offensive character. For a "properly balanced" air arm, the chief believed, air service should comprise 20 per cent of the total available strength and air force 80 per cent. The prevailing disposition of strength—one group each of pursuit, observation, and bombardment—was not, he said, an air arm of proper composition.[49] The composition of the air arm proposed by the new chief represented a concept of doctrine breaking sharply with that embedded in military thinking. This new concept was, in fact, little more than a reversion to the doctrine implicit in the programs of the Air Service, AEF, before the final 202 program which reflected the probabilities of production rather than consciously determined policy. In urging the reversion, the Air Service was doing little more than suggesting a return to the trends already begun in Europe. But it was too late; the damage had been done. The "accepted" lessons of the war rested on the 202 program in which aviation for observation predominated. Having failed to establish an organization to analyze the experience of the war objectively and systematically, the Air Service saddled itself with an inadequate doctrine which it found hard to

48. Further indication of the limited air doctrine taught in military schools is available in the course materials used by the Army War College during the period immediately following World War I. A survey of the texts available in the National War College Library revealed that analytical studies of doctrine, apart from résumés for the staff schools of all arms and services, were literally nonexistent until the mid-thirties.

49. Report of Chief of Air Service, *Annual Reports of the War Department,* 1921, *1,* 185. The language used in this report, especially such phraseology as "air service" and "air force," is almost a direct translation from Douhet. See G. Douhet, *The Command of the Air,* tr. Dino Ferrari (New York, Coward-McCann, 1942).

escape.[50] Once that doctrine, no matter how faulty, became a part of the heritage of World War I, it proved difficult to alter. The airmen, forced to swallow the pill of their own failures, had themselves chiefly to blame. As a consequence, they were to spend a generation struggling to correct the deficiency.

50. The separate-air-force controversy which goes beyond the scope of this study was brought about, it would appear, in an effort to rectify the neglect of the air arm itself in formulating doctrine.

## Chapter XI. Summary and Conclusion

EVEN THE MOST cursory survey of military history substantiates the premise that superior weapons give their users an advantage favoring victory. A somewhat closer study of military history shows that new and more effective weapons have generally been adopted only slowly in spite of their obvious advantages. Since the character of contemporary weapons is such that their production as well as their use can dislocate whole economies, it is probably not too much to suggest that the survival of entire cultures may hinge upon an ability to perfect superior weapons and exploit them fully. Survival itself, then, appears to depend on speed in both the development and the utilization of weapons.

The air weapon, from its inception down through the period of World War I, offers a case study illustrating the problem of weapons as a whole. In 1914 the airplane was an untried innovation replete with unweighed potential. World War I presented the United States with an opportunity to exploit that innovation. This study has been undertaken on the assumption that a close analysis of how the United States made use of the air weapon might reveal significant lessons useful in meeting the problem of weapons in the future.

Exactly what are the lessons to be derived from the experience of the United States with the air weapon during World War I? These lessons are much the same as those which might have been derived equally well from the Civil War or, for that matter, from any other war. As was true of former conflicts, World War I emphasized the necessity for a conscious recognition of the need for both superior weapons and doctrines to ensure maximum exploitation of their full potential. As a corollary to these two requirements, the war pointed up the need for administrative agencies to ensure their fulfillment once they have been recognized as requirements. The experience of the war showed clearly that wherever military authorities failed to emphasize the need for better weapons rather than more weapons they suffered serious disadvantage. Aerial warfare along the front proved that an enemy with fewer but superior weapons was fully capable of containing an opposing force with an impressive numerical predominance. Quality paid better dividends than quantity.

The experience of the war also demonstrated that where military authorities failed to formulate a doctrine to exploit each innovation in weapons to the utmost they suffered further disadvantage. The example of bombing aircraft presented an outstanding instance of this neglect. Not only did the military authorities fail to get bombers into production for immediate use during the war, but as a result of their neglect of doctrine, the air arm acquired no body of experience from which to derive an acceptable concept of aerial warfare. Lacking such a concept or doctrine, the air arm had little to give authoritative direction to development of aircraft for the future.

Finally, the experience of the war showed that the failure to emphasize better weapons rather than more weapons and the failure to attach sufficient importance to the formulation of doctrine issued directly from inadequate organization. The war revealed that adequate organization fell into two general categories. The first of these involved organizations for information, that is, agencies for objective, systematic compilation (at all levels of operations) of facts, and of facts, indeed, about aerial warfare and doctrines of air power both foreign and domestic, facts regarding tactical developments to serve as a basis for countermeasures, facts about technical developments, about the result of proof tests, and about scientific findings for possible application to weapons. The second of the two categories of organization involved means for making decisions. The war showed the necessity of organizations at all echelons for making authoritative decisions based upon information systematically, objectively, and continuously accumulated by responsible and effective organizations especially created to gather data. At the same time, the events of the war showed that decisions based upon opinion, memory, a limited range of personal experience, or emotional bias led only to failure. These were the lessons of World War I, but did the air arm learn them?

The evidence here presented leads to the conclusion that the postwar Air Service made use of only a relatively small portion of the experience of the war regarding the problem of weapons. For want of a full appreciation of the need for retaining every last possible lesson of experience, the Air Service lost or abandoned many vital policies, procedures, methods, and practices which had to be relearned by painful practice in the subsequent years of peace and war. But with specific regard to the problem of weapons, there were significant lessons learned as well as lessons lost.

The postwar Air Service did learn conclusively that better weapons were more desirable than more weapons, but unfortunately some of the best administrative mechanisms devised during the war to secure superiority of weapons were abandoned or ignored in peace. The Air Service manifestly did not learn from wartime experience the critical importance of systematic formulation of doctrine as a step essential to successful development of air weapons. In consequence, the growth of the air arm in peacetime suffered a significant handicap.

The available evidence shows that after the war the Air Service learned the importance of organization for decision and established channels of command for unified, decisive, and authoritative action in contrast to the dispersed, ill-defined, and overlapping channels which existed during the war. On the other hand, the record suggests that the Air Service did not entirely learn the lessons of the war regarding the importance of organization for information, for the air arm in the postwar period was certainly deficient in organizations to secure and study information about foreign concepts of air doctrine. There were no organizations in the air arm for analyzing critically as well as objectively all aspects of the heritage of wartime experience. Similarly, the air arm lacked an organization for continued liaison with science. In sum, these lessons learned and unlearned are the measure in which the United States made use of that hitherto untried innovation, the airplane.

The student of military history should be able to draw a number of significant lessons from the findings of this study which may bear upon the larger problem of weapons as a whole. The first chapter of this inquiry described the English longbow, an innovation in weapons which sat on the kingdom's back doorstep for 250 years before it was finally exploited to overcome the prevailing French superiority in numbers and in armored, mounted knights. Those 250 years may appear to have been an excessively long delay in the exploitation of a weapon at once so significant and capable of giving such an obvious advantage. In a vastly accelerated age, it may be that the United States did no better in waiting some 20-odd years before exploiting the air weapon to the utmost.

Many of the events in this study are a generation old. Subsequent history, particularly that of World War II, tends to obscure the record of the former war. Nevertheless, the mistakes and lessons of World War I still repay analysis. While it is undoubtedly true that the nation has learned much from the experience of two world wars, many lessons yet remain to be uncovered if superiority in weapons is to be maintained. In emphasizing the importance of method in the development of weapons it is well to recall that as late as 1939 the navy gave the atom bomb the polite brush-off. Furthermore, as late as 1940 when the Germans were reported to be diligently in quest of atomic power for military purposes, the United States placed only $6,000 in federal funds behind atomic studies in this country.[1]

One way to measure the value of this study, would be to read "guided missile" or any other recent and novel weapon for "aircraft" of vintage 1918. Thoughtful citizens, layman and specialist, politician and staff officer, would do well to ask and ask again if the military establishment of the United States is sufficiently well organized to develop and exploit to the utmost these newest weapons on the horizon. Failing here, the nation will repeat the sorry pattern of the air weapon, wastefully groping forward with each innovation. To exist in a warring world the nation must pick winning weapons; if military analysts will distill every possible lesson from the history of two world wars such weapons will be easier to find and the odds on national survival will go up.

1. H. D. Smyth, *A General Account of the Development of Methods of Using Atomic Energy for Military Purposes under the Auspices of the United States Government, 1940–1945* (Washington, GPO, 1945), pp. 32, 33.

*Bibliographical Note*

## INTRODUCTION

THE CENTRAL FOCUS of interest in this inquiry is the exploitation of the aerial weapon by the United States during the period of World War I. By far the greater portion of the source materials consulted were the official records of the period in question. These included the general correspondence or office files of interested agencies, official publications of annual reports, statistical summaries, and the like, as well as first narrative drafts of studies prepared by historical units at various echelons of the War Department. Since the nonspecialist and nonmilitary reader requires a rather extensive setting to place the problem in its proper context, the early chapters of this study are primarily introductory. As such, they are quite consciously based for the most part on frankly secondary materials.

Chapter 1 leans heavily on published materials, most of which are readily available. Chapter 2, dealing with the origin of the interest of the War Department in the air weapon, is but a brief sketch for background. The period merits a much more detailed study than the scope of this book permits, but since this chapter is primarily concerned with introducing the general reader to the problem, it appeared sufficient to base its contents almost entirely upon such source materials as reports of the War Department, semiofficial histories, and critical monographs prepared during World War II by members of the historical staff of the AAF. Chapters 3 through 10, the body of the study, are based chiefly on official documents nearly contemporary with the problems discussed. Memoirs and similar works have occasionally been used but always with caution. The disparity between recollections recorded in retrospect and the facts available in the files of the War Department leads one to the emphatic conclusion that the human memory is highly fallible as a source of historical evidence.

## I. BIBLIOGRAPHIC AIDS

Since this study embraces two fields, the use of the aerial weapon by the United States during World War I and the broader problem of development of weapons as a whole, there are two distinct sets of bibliographic aids which bear on the subject. Bibliographies dealing with aviation during World War I are, at best, disappointing. General bibliographies dealing with the war at large occasionally devote a brief section to aviation, but the titles included seldom go far beyond a dozen or so items, mostly popular publications.

The best single guide to the most important publications on American aviation in World War I is a Congressional publication, *Pioneer Aviators*, Hearings before the House Military Affairs Committee on HR 11273, 70 Cong. 1 Sess., 3 April 1928, pp. 18–19. Though brief and by no means complete, the titles here listed embrace most of the available published materials including Congressional publications, which are of particular importance.

Probably the most extensive bibliography of periodical literature on aviation in World War I is the NACA *Bibliography of Aeronautics*. The volume covering the period 1917–19 (Washington, GPO, 1923) is of particular value, although editions of successive years have entries of interest.

The most fruitful approach to the problem of bibliography made in planning research for this study came from a survey of the bibliographies contained in the monographs prepared by the Air Historical Office of the AAF in connection with the historical program of World War II. (See discussion of individual monographs below.)

A guide to unpublished materials on aviation in World War I is to be found in the National War College Library. This compilation, "Aeronautical Information," a mimeographed list of technical reports, translations, intelligence reports, etc., was issued periodically by the Executive Section of the Division of Military Aeronautics. Unfortunately, the first list did not appear until September 1918, and the last was issued in June 1919. Many of the documents listed are no longer obtainable, but the lists provide something of an index to the types of materials for which one might search. Although the title page describes the lists as abstracts, they are in fact more properly calendars. The lists are catalogued in the National War College Library under Z5064, U53, Nos. 1–40, 1918–19.

Taken all together, the bibliographies cited above offer a most unsatisfactory set of guides to the problem of this inquiry. The absence of formal guides is probably inevitable since the special field of interest here considered is still virtually untilled. If this is true of the aerial weapon during World War I, it is even more applicable in the case of the larger problem of the development of weapons in general. Bibliographies in this field are virtually non-existent, and those few which do exist are largely concerned with books on the manufacturing of weapons, such as cannon casting or gunsmithing, rather than with the administrative problems of selecting, evaluating, and developing weapons. In the absence of any specialized guides, probably the best approach to the problem of weapons is to consult the bibliographies and citations in the writings of the few authors, such as Brig. Gen. J. F. C. Fuller, who have shown a particular interest in the field.

No bibliography dealing with source materials in the War Department would be complete without mention of the manual on filing published by the Adjutant General of the army. Anyone contemplating research in records of the War Department would do well to invest a half-hour's time in studying the volume entitled *War Department Correspondence File*, first published in Washington during 1917 and revised periodically since. This manual describes the filing system, a decimal derivative, used by the department. Whatever may have been the failures and misapplications of this filing system and however great may have been the abuses perpetrated by the various organizations using the filing principles laid down, a familiarity with the manual cannot help but assist in the work of research.

## II. MANUSCRIPT AND OTHER UNPUBLISHED SOURCE MATERIALS

The unpublished materials used in the preparation of this study were drawn from a large number of collections including those of the National Archives, the Air Force Central Files, the Air Historical Office, the National War College Library, the Library of the Industrial College of the Armed Forces, and the Wright Field (Air Materiel Command, Dayton, Ohio) Historical Office Files. Materials used from each of these collections are described in some detail below.

### National Archives

Of all the archives explored, probably none was more useful than the National Archives, which houses a whole series of record collections pertinent to this study. These several separate collections fall within two different official classifications within the larger framework of the system of classification used by the National Archives. Record group 18 contains materials pertaining to the Air Service and the Bureau of Aircraft Production in the United States. Record group 120 contains materials pertaining to the Air Service, AEF. During World War II the former was utilized to a limited extent by the handful of officers who were aware of its existence. The latter, containing the AEF materials, has apparently remained virtually untouched since immediately after World War I. Therefore, because these collections of documents represent a vein of unworked materials, the detailed description undertaken below is somewhat more extensive than might otherwise seem necessary.

Record group 18 in the National Archives contains at least seven different sets of files which were of particular interest in the preparation of this study:

1. *The Bureau of Aircraft Production Historical File.* This collection, which consists of more than 30 file boxes, houses the working papers accumulated during 1918 and 1919 by a historical staff in the Air Service which was preparing a record of the BAP during the war. The labels on the boxes, as in the case of most items in record group 18, are misleading. The caption, as given, is "Army Air Forces, Bureau of Aircraft Production, 1918–1920." The use of AAF here is of course erroneous, and the contents of the files cover the period 1917–19 rather than 1918–20 as indicated. (All the other file collections of record group 18 in the National Archives bear the misleading caption of the AAF.) This collection of files contains voluminous compilations of statistics regarding aircraft production, copies of correspondence concerning policy on the development of aircraft, and a large amount of information on organization. Scattered throughout the files are typescript copies in rough draft of chapters prepared for a history of aircraft production during World War I. These materials were exceptionally useful in writing this study and, in addition, contained many leads to materials in other collections.

The caliber of the draft chapters of the history of the BAP varies markedly. Some, although substantially accurate in fact, are almost incredibly naïve. Many of the chapters appear to have been written by persons without a comprehensive grasp of the larger problems encountered in developing the aerial weapon. At least 18 different writers participated in the project and possibly as many more research workers. The writers appear to range in ability and historical objectivity from a few well-trained historians, cautious and wary in their treatment of the evidence available, to those with a standard of writing best characterized as journalistic. The evidence of the files indicates that the officer supervising the project, Capt. J. L. Ingoldsby, Chief of the Air Service Historical Section, was a painstaking workman setting high standards of accuracy in detail though willing to accept narrative accounts in a large measure devoid of evaluation and analysis.

The Historical File of the BAP is of interest as an example of an unusual historical technique. Confronted with the task of writing a history of aircraft production, the historical staff of the Air Service established a set of files using the standard decimal system of the War Department. Into this framework the staff apparently poured all manner of historical materials—copies of correspondence, clippings, Congressional hearings, reports from the War Department, and the like. Then, using this as a background, research workers in special fields pursued specific lines of interest, compiling individual dossiers of information for each chapter in the projected publication. These folders of "foundation data," as they were called, contain chronological outlines of the problem in question, true copies of important pieces of correspondence, statistical tables, reports, and illustrations. The chapters written from these compilations appear to have been the work of teams of writers and research workers rather than individuals. The marginal notes, buck slips, and editorial changes entered in Captain Ingoldsby's hand provide an additional basis for appraisal of the documents compiled.

The history of the BAP, which was finally completed during 1919 and 1920, consisted of more than 2,000 typescript pages. Unfortunately this manuscript never reached publication. One copy of the draft in typescript, the original, is believed to have been lost in the fire which destroyed so many records of the Air Service at Bolling Field, Anacostia, District of Columbia, during the early twenties. Another copy, consisting of several bulky volumes of carbon second sheets, was "lost in the files" for more than 20 years. It was recovered during World War II from an obscure corner in a Detroit office building by the Wright Field

Historical Officer, too late to be of real value to staff officers in this war. This second or carbon copy is now on file at the Wright Field Historical Office, and a microfilm "edition" of it is on file in the Library of the Industrial College of the Armed Forces in Washington. The utility of this manuscript is considerably impaired by a lack of documentation, apart from occasional internal references in the text itself. Although the history of the BAP actually was based on official records, as the working papers in the National Archives reveal, a careful historian working only with the microfilm or typescript copies would have no such assurance.

The working papers of the historical project of the BAP in the National Archives which were of exceptional value in preparing this study are listed below by box number and file number. These papers should be considered virtually indispensable to anyone studying the development of aircraft during World War I because they represent an all-inclusive survey or cross section of all the agencies involved.

Box 4: see especially files 320.3, Reorganization of the Air Service, and 321.9, Functions of the Air Service, as well as 321.9, Air Service Program.

Box 5: see 321.9, Requirements Division.

Box 6: see especially 311.2, Ribot Cable. This file contains many items of particular interest regarding Ribot's fateful cable.

Box 7: see 314.7, Bureau of Aircraft Production History, General, which contains many chapters in rough draft of the unpublished history.

Box 8: see especially 333.5, Investigations, General. This file contains a chapter by one J. A. Beck dealing with the several investigations of the aircraft program. Because of its controversial nature and frankly critical contents, it was to have been deleted from the published version.

Box 9: see 334.7, Aircraft Production Board. This file contains an index to the central files of the board, a useful research tool. See also 334.7, Joint Army-Navy Technical Board, and 334.7, National Research Council History. Under 334.8, Overseas Missions, a draft chapter on the history of the BAP entitled "The Formulation and Distribution of Programs" is misfiled. This was particularly helpful in the preparation of this study, although the text is subject to criticism as containing many unwarranted assertions, the tables of statistics presented do check with other verifiable materials.

Box 10: under file number 334.8 see especially the files on the

Bolling Mission, the Spaulding Mission, the Waldon Mission, and the Lockhart Mission. Each of these files and the draft manuscripts of chapters they contained were of particular interest to this study. Under 336.91 see British War Mission, and Foreign Missions, the latter rather scanty.

Box 20: see 452.1, Airplanes along the Western Front, containing Lt. Col. V. E. Clark's postwar summary of the problem of the air weapon, an unusually shrewd appraisal. This box contains a file, also under 452.1, on the Bristol Aircraft.

Box 21: see 452.1, Caproni Contract, and 452.1, Caproni History.

Box 23: see 452.1, Handley-Page.

Box 31: see 471.62, Aerial Torpedo.

2. *Bureau of Aircraft Production, Executive Office Files.* National Archives record group 18 contains a number of other collections of files of interest to this study. Next in importance to the Historical File are the materials of the Executive Office of the BAP. As the coordinating office of the BAP, the Executive Office's papers, like those of the Historical File, present more of a cross section than one might expect to find in any other agency at the operating level. These papers include general correspondence, staff studies on policy, interoffice memoranda, buck slips, and the like, many of them entertaining as well as significant. Items of particular interest are listed below.

Box 6: see 026.4, Bureau of Aircraft Production, 026.4, Scientific Research, and 026.4, Air Service, AEF.

Box 65: see 334.8, Aircraft Board.

Box 67: see 334.8 entitled Bechereau Mission, which contains data on the SPAD Mission.

Box 81: see 402.1, Technical Data, one of the most useful files in the entire collection of the Executive Office, with many insights into the complexity of the problem of liaison for technical information.

In addition to the BAP Historical files and the files of the Executive Office, record group 18 in the National Archives contains five other file collections which, though extensive, were of relatively less value in preparing this particular study.

3. *The Files of the Administrative Division of the Division of Military Aeronautics.* These files, the DMA counterpart of those from the Executive Office of the BAP, contain a few items of great value to this study, but most of the contents deals with postwar problems. The inclusive label date, 1918–20, is applicable

here. Some of the more helpful items regarding the problem of developing the aerial weapon are listed below.

Box 3: 321.9, Science and Research Division, contains some correspondence which rounds out the materials in the files of the BAP.

Box 12: 400.112, Specifications.

Box 14: 452.1, Airplanes 1919–1920, contains data on the postwar establishment.

Box 16: 461, Libraries, contains materials on the problem of technical information.

Box 20:835, Executive, contains items regarding organizational policy.

In this file collection, as in all the others included in National Archives record group 18, the frequency of misfiling impedes research. The labels are no certain indication of contents. While the decimal system of the War Department gives an approximately standard framework to all the files, one cannot select a file number and expect to find under it all those materials originally filed there. A generation of moving about from one warehouse to another has transposed whole segments of records. Individual folders contain as many as four different decimal files. These materials have not yet been worked over by the staff of the National Archives, and the research worker should not be surprised to find a heterogeneous collection of dime novels, newspapers, and other such materials hastily cached by long-forgotten file clerks at the unexpected appearance of a supervisor. In short, there is no alternative to the slow process of leafing through each individual file.

4. *The Bureau of Aircraft Production Miscellaneous Historical File.* This collection of less than a dozen boxes contains only one box of worth-while materials. Although the archival records designate these files as the Miscellaneous History Files of the BAP, the shelf boxes are entirely unlabeled, making the game of research less scientific but infinitely more interesting.

Box 1 contains a number of very useful items on organization, including files 167.2, Peace Organization; 321.9, Organization of DMA, with many extraneous items not at all connected with DMA; 321.9, Organization of the Air Service, and 321.9, Air Service, Training, History and Organization, AEF; and 026, BAP. This same box also contains two items not filed decimally. One of these, a lecture by Maj. F. P. Lahm given at West Point 13 February 1920, is entitled "History and Development of the Air Service," a rather pretentious heading for such a brief sketch. Another pamphlet, with the same caption on the folder bears the title "History of the U. S. Army Air Service 1862–1920" with the notation "Printed (mimeographed) on order of The Adjutant General for publication by Century Co.," 1 October 1920. Apparently the project to publish fell through because no such volume seems to have appeared.

5. *The Finance Advisory Board File.* This file, labeled Air Service 1918–1926, contains a series of case folders, serially numbered, each case being a staff study made by the board. The label "Finance Advisory Board" is misleading. The board may have been so called for a brief period, but during most of its existence it was called simply the Advisory Board. In reality it was a rudimentary staff to the chief of the postwar Air Service. The cases or staff studies contained in this collection vividly illustrate the organizational weaknesses of the new air arm. The following cases were useful: 179, 219, 291, and 310. Taken collectively, the series represents an excellent object lesson in organization and operating methods of a staff.

6. *Office of Chief, Air Service, Files.* This file collection is erroneously labeled "Balloon School." Actually it contains the files of the Executive Office of the Air Service, Office of the Chief of Air Service. Largely concerned with the period 1919–26, these files contain only a few items of interest to this study. See especially Box 170, with charts on the organization of the Division of Military Aeronautics.

7. *AEF Cable File.* The seventh collection of files in National Archives record group 18 of particular interest to this study is radically different in character from all the other collections. This file contains copies of cables from the AEF to the United States. Several sets of cables fall within this group. These include Pershing's cables to the War Department (regarding aviation), Sims' cables to the Navy Department (regarding aviation), and Bolling's cables to the War and State departments. The collection contains messages originating in both London and Paris.

The cables embraced in this collection appear to be "action copies" rather than code-room originals. To historians not familiar with military practice a word of explanation may be helpful. Although the best historical tradition teaches that "originals" are more authentic than "copies," the file of cables presents a special case. The original typescript sheets from the code room consist of long messages to the War Department as a whole. As they were decoded, sections of these blocked messages were broken off and

sent to the chief of service concerned, or the Chief Signal Officer
in the case of aviation in 1917. Such copies, in the hands of individual chiefs of service, became "action copies," the bases upon
which subsequent actions were taken. Thus, the "copies" became,
in some ways, more important than or at least equally as important as the "originals." The cables of the AEF in record
group 18 appear to have been "action copies" for the Signal
Corps. The fact that they bear marginal notations, queries, and
interlinear additions supports this view.

To understand something of the difficulty surrounding the use
of transatlantic cables, one must appreciate their appearance on
receipt. Written entirely in capital letters with a minimum of
punctuation confined to spelled-out instructions, the cables were,
at best, confusing. The difficulties of transmission were further
increased by an acute shortage of trained cablemen and code
clerks. Added to these difficulties was the confusion resulting from
varied routings of messages; some were sent direct, some through
the military attachés in London or Paris, and some through the
State Department.

The cables of the AEF on aviation in the National Archives are
still tied in corded bundles which are hard to use. All are arranged
chronologically. Some are grouped by subject, e.g. "SPAD Aircraft" or "Liberty Engine"; others, mostly questions of policy,
are grouped under the heading "Pershing Cables," although many
are of course signed by subordinates. Postwar Congressional hearings and contemporary periodical literature show rather amusing
evidence of surprise at learning that all cables signed "Pershing"
were not really seen by him. The cables from the AEF on aviation,
by whomever signed, consist of a jumbled mass of documents in
which personnel matters, personal trivia, highly technical engineering data, as well as broad questions of policy are all run together.
This jumbled mass of information was sometimes additionally
garbled by the language of transmission, which should give the
reader an appreciation of some of the difficulties lying behind the
program for aviation during World War I. Any attempt to assess
the transatlantic relations of the day should include this factor
of confusion in exchanges by cable. These messages were highly
useful in the preparation of this study even though copies of almost all significant cables are to be found scattered throughout
the files of the operating agencies. An additional word of warning
may be in order: many of the cables are meaningless, or relatively

so, unless read in conjunction with letters sent by slower transportation.

Of the two major collections in the National Archives of materials from World War I pertaining to aviation, record group 18 was probably the most significant in shaping this study. Nevertheless, materials classified in record group 120, dealing with organizational records of the Air Service, AEF, were useful in supplementing the point of view of "supply" with the point of view of "operations." In the broad range of record group 120, two different file collections were consulted, the Air Service Historical Records and the Air Service History itself.

1. *World War I Organization Records, Air Service Historical Records.* These files, more than 2,000 shelf boxes, consist primarily of the working papers of the Air Service, AEF, historical project begun in France during the closing months of the war and continued in the United States during 1919–20. Included in these files are the papers used in the preparation of the history of the Air Service, AEF (see below), the papers used in preparing the final report of the Chief, Air Service, AEF, and those used in drafting the articles published in the postwar *Air Service Information Circular.* The collection includes chapters in draft form and, in some cases, materials used in compiling those chapters for the unfinished history of the Air Service. The files also included some histories of subordinate echelons. The correspondence and administrative files of the historical project itself provide revealing indications of the methods used in conducting research and compiling the reports contained in the collection. Like that of the BAP, the historical project of the Air Service, AEF, used the decimal filing system of the War Department as a master framework for collecting data over the entire range of subjects involved.

The project appears to have been under the supervision of Col. E. S. Gorrell, one of the boy colonels of World War I (he was 27 at the end of the war). The career of Colonel Gorrell presents a paradox. While a member of the Bolling Mission in 1917 he advocated mass production of bombers by the United States. Then, during the latter part of 1918 when he served as Chief of Staff for the Air Service, AEF, and later as head of the historical project, the colonel advocated the primary importance of "air service" over "air force," putting first priority on aircraft for observation. Finally, writing and speaking in 1940 as a civilian official in an association to foster commercial air transport, Gorrell once again favored strategic bombardment. Since this individual officer played a critical role all during the period discussed in this study, a narrower investigation of his shifting opinions might prove highly meaningful.

Items of particular interest in the Historical Records of the Air Service, AEF, are listed below.

Box 1 contains copy in rough draft and data on the research used in compilation of the final report of the Chief, Air Service, AEF. This box also contains data for the report of the assistant chief and "Tactical History of American Day Bombardment Aviation," a brief draft in typescript. Perhaps the most important item of Box 1 is a decimal file index or guide which provides at least a rough map of the relatively unexplored materials in the 2,000 boxes of this collection.

Box 2 contains data collected by Col. E. S. Gorrell referring to the early activities of the Air Service, AEF, i.e., to the period of the Bolling Mission.

Box 3 contains excerpts from Gorrell's testimony before Congressional bodies during the postwar period.

Box 300 contains versions in rough draft of several chapters (iii, iv, v of Vol. *1*) of the history of the Air Service, AEF, which never reached publication. Marginal notes, buck slips, and other file contents show that some of the chapters were prepared with considerable care, quotations checked for accuracy, documentation checked, etc. However, from such materials as are available, it would appear that as a whole the historical project of the Air Service, AEF, was neither comprehensive nor critical.

Box 317 contains printed copies of the 100-odd *Information Circulars* published by the Air Service to consolidate the lessons of the war. Although specifically declared to be "unofficial" (i.e., lacking formal approval of the General Staff), these publications mirror the prevailing opinion in the formal reports of the Air Service, AEF, and the postwar Air Service in the United States.

Box 512 contains, under file 452.1, a copy of the report of the Statistics Branch of the General Staff, "Distribution of Service Planes as Related to Military Policy," a document of particular concern to the central thesis of this study.

2. *Air Service, AEF, History.* This collection of 60-odd volumes in typescript of varying format is irregularly titled. It is commonly called "Gorrell's History" or by the designation in the heading above. Only one typescript copy is available. As the final

product of the historical project of the Air Service, AEF, this collection is a curious hodgepodge. It appears to consist of an uncritical assembly of letters from commanders at various echelons, histories of units, and organizational studies. Emphasis appears to rest heavily on form rather than upon operations or functions. Documentation is occasional and casual. The volumes entitled "Lessons Learned" belie the promise of their title; typical of the whole series, they are unsystematic, not comprehensive, and inconclusive. In the absence of anything else of a comparable nature, however, these histories are valuable and useful. Nevertheless, considered collectively, they represent a disconcertingly slim heritage from the aerial operations of World War I. In the preparation of this study the histories were used sparingly in favor of the source materials contained in the working papers from which they were compiled.

## AIR FORCE CENTRAL FILES

The Army Air Forces Central Files, inactive section, contain some materials of use in this study. These files, the unclassified central files of the Air Service, Air Corps, and the Army Air Forces Headquarters, are officially considered to date from 1919 to 1945, but there are many papers preceding the initial date. Portions of this collection remain in the custody of the Adjutant General's Office, Records Branch (Air Force Section), in a warehouse near the Pentagon, but the bulk of the material has been transferred to the National Archives since the completion of this study. Altogether the collection contains more than 1,900 individual folders of correspondence, staff reports, interoffice memoranda, and papers of a similar nature. Only a very small portion of this material dates back to the period of this study.

The Air Force Central Files use the decimal system of the War Department down to the year 1942. The following file numbers offer a guide to pertinent materials:

041.2, Military Attachés
059, Statistics
334.7, Advisory Boards
334.7, Airplane Boards
334.7, Army-Navy Joint Boards
334.7, Board on Organization of the Air Service in Peace
334.8, National Advisory Committee for Aeronautics
334.8, National Research Council
350.05, Collection of Military Information

350.051, Dissemination of Military Information
360.02, Foreign Aviation
385, Methods and Manner of Conducting War
400.112, Tests and Experiments
452.1, Experimental Airplanes

### AAF HISTORICAL OFFICE

During World War II the AAF established an elaborate historical project which survived into the peace somewhat more successfully than its predecessor of World War I. Since the name of this organization has changed with the seasons, throughout this study the documents in its files have been cited AAF Archive. This historical project of World War II encompasses an impressive program of histories of units at all echelons and specialized monographs of operations at headquarters as well as comprehensive studies on problems of policy. The Air Historical Group, as it is currently called, has some 6,000-odd unit histories in its files with an estimated 400,000 supporting documents. Of approximately 100 monographs projected, more than 70 have been completed. The catalogue of the Air Historical unit offers a useful bibliographical tool for research in military aviation. Some of the monographs completed by the office during the war and reproduced in five or six typescript copies were of value in preparing the background of this study. Well documented and written, for the most part, by trained historians, these monographs represent a significant contribution to the slender literature on policy regarding the development of aircraft. Although these monographs are not strictly source materials, for convenience they are included here with the discussion of the source materials of the archive in which they are to be found. Moreover, many of the monographs contain appendices of significance as basic sources, and in preparing this study the use made of them was primarily for these source materials rather than for the conclusions expressed in the monographs. The following were most frequently consulted:

No. 25, "Organization of Military Aeronautics; 1907–1935," issued December 1944; contains appendices with basic legislation; useful for reference.

No. 39, "Legislation Relating to the Air Corps Personnel and Training Programs; 1907–1939," issued December 1945.

No. 44, "Evolution of the Liaison-Type Airplane; 1917–1944," issued May 1946; presents the case history of one weapon as illustrative of the broader problem of aviation as a whole.

No. 50, "Material Research and Development in the Army Air Arm; 1914–1945," issued November 1946; an exceptionally valuable study analyzing many of the factors behind the development of aircraft; contains a useful section on the NACA.

No. 54, "The Development of Aircraft Gun Turrets in the AAF; 1917–1944," issued May 1947; a case study similar in scope and purpose to No. 44, cited above.

In addition to the monographs prepared by the Air Historical Group, the files of the Historical Office contain a large number of records collected while the monographs were being written. These records include such items as organizational manuals of DMA and the early Air Service, organization charts, staff studies, and the like.

## WRIGHT FIELD HISTORICAL OFFICE FILES
### (*Air Materiel Command, Dayton, Ohio*)

Wright Field, or the headquarters of the Air Materiel Command, was the center of experimental engineering for the Air Forces. For this reason the files of the Historical Office contain many items of interest in connection with the problem of development of airplanes. During World War II the historical program at Wright Field was coordinated with the historical program of Headquarters, AAF, in Washington, and the remarks above concerning the collection of records at the Washington office apply generally to those at the office at Wright Field. In this study the files at Wright Field proved exceptionally useful with regard to the Engineering Division and its initial problems in the transition from war to peace. The most important single item at Wright Field pertaining to World War I is the multivolume copy in typescript of the history of the BAP, the only known complete copy in existence, excluding, of course, microfilm reproductions of this copy.

## NATIONAL WAR COLLEGE LIBRARY

The National War College Library should not be overlooked as the repository of an unusual collection of documents. It also houses an extensive collection of publications on military subjects. Many military records of transient official interest, short-term periodicals, reports of units long since deactivated, and other similar fugitives have been accessioned in the War College Library where they are, in some cases, unique items. Included in this body of materials, largely typescripts or photostats, are numerous translations from foreign military periodicals, copies of lectures delivered at the War College, reports by students on special topics, statistical reports of the General Staff, and operational reports from units in the field with the AEF and other organizations. Lectures and papers by students were useful in documenting the character of military thought on air doctrine during the postwar period, and statistical studies of the General Staff were of particular value in appraising methods used by the staff in formulating doctrine. Outstanding among those documents in this collection which were of importance in the preparation of this study were the reports in the periodical series published by the Air Service, AEF. Variously styled "Air Service Activities," "Weekly Reports," or "Progress of Air Service Activities" and issued first by the Coordination Staff, Air Service, AEF, then by the Executive Section, this series of 31 reports ran from October 1918 to May 1919. There were 31 volumes in all, the final volume containing an index. The whole collection provides a compendium of factual data, statistics, and the like on activities of the air arm. The volume of Armistice week is a particularly useful review.

The collection of documents at the War College also contains one of the analytical reports on air operations issued by the Statistics Branch of the General Staff, "Statistical Analysis of Aerial Bombardment," Report No. 110, 7 November 1918. Unfortunately, only scattered reports from this series are available. There is, apparently, no complete collection extant. For that matter, there is no evidence available to indicate that the projected series was ever completed.

### LIBRARY OF THE INDUSTRIAL COLLEGE OF THE ARMED FORCES

Like the National War College Library the Industrial College Library contains a collection of documents which may be interesting to the student of development of the air weapon. But since comparatively few items of special value in the preparation of this study were found there, the collection requires little more than mention.

# III. PUBLISHED MATERIALS

Published materials utilized in this study fall into the two customary groups—primary source materials, such as official reports of the War Department and Congressional reports and hearings, on the one hand, and unofficial, more or less secondary materials on the other hand.

## PUBLICATIONS OF THE WAR DEPARTMENT

1. *Annual Reports of the War Department*. These reports, including the report of the secretary, the Chief of Staff, and the reports of technical services such as the Signal Corps, were consulted frequently in the preparation of this study. Reports most frequently employed were those covering the years 1904 to 1922. Anyone utilizing annual reports of the War Department should be familiar with the method traditionally used in compiling them. The chiefs of divisions in subordinate echelons are called upon for contributions covering the activities of their organizations. The sum total of these reports is then edited at the echelon issuing the report. The final result is thus frequently somewhat uneven as to both form and content. It often happens, when the highest echelon limits its editing to syntax rather than content in terms of policy, that the report as published contains views more nearly those of the chiefs of the operating divisions than of the chief of the higher echelon who passively accepted and approved the reports sent forward to him. It is this method of compilation which occasionally results in annual reports of the War Department containing special pleas of a disproportionate character for personnel, funds, or recognition of relative importance for subordinate units.

2. *Special Reports of the War Department*. In this group of materials are included all publications appearing within the department but not necessarily bearing the imprint of the department. These publications, for the most part consisting of items issued by the chiefs of Arms and Services within the War Department, represent some of the most significant materials used in this study. They embrace intelligence reports, studies on policy and final reports of operations, statistical summaries, and related materials. Those most frequently consulted or of particular value are discussed below.

In the general field of method in the development of weapons broadly considered, probably no single publication was more useful or more entertaining than Brig. Gen. S. V. Benet, ed., *A Collection of Annual Reports and Other Important Papers Relating to the Ordnance Department* (Washington, GPO, Vol. *1*, 1878, Vol. *2*, 1880, Vols. *3*, *4*, 1890; title varies slightly in successive volumes). Edited by a onetime Chief of Ordnance, this little-known collection of documents relating to the Ordnance Department is even more valuable than its title might indicate. Historians with problems ranging far beyond the confines of ordnance could utilize these documents, which reflect interests in every sphere of public life, social, economic, and political. An extremely detailed index greatly enhances the value of the volumes. Correspondence reproduced on p. 478 and the pages following in Vol. *3* provides useful bibliographical data regarding the compilation of records of the Ordnance Department.

In the narrower field of the air weapon in particular, there are about a dozen publications emanating from various offices within the War Department which are of special interest. An early and unusual statement of policy regarding the air weapon is to be found in *Military Aviation*, House Document No. 718, 62 Cong. 2 Sess., 26 April 1912. This 80-page document contains a letter by the Secretary of War transmitting data on aviation requested by the House of Representatives. It includes information on the status of military aviation in Europe, the existing establishment for aviation in the Signal Corps, and programs recommended for future expansion. Particularly interesting is the nonconcurrence of Secretary H. L. Stimson in proposed expansions of personnel for the air arm.

Brig. Gen. G. P. Scriven, *The Service of Information* (Washington, GPO, 1915), published as Circular No. 8 of the Office of the Chief Signal Officer, contains an early statement of doctrine. Although nominally attributed to the Chief of the Signal Corps, this item probably represents the combined efforts of several officers. It contains some elaboration on doctrine as formulated by the signal chief before Congressional hearings but adds little or nothing to the basic policies earlier enunciated. Another document of this character is *Military Aviation* (Washington, GPO, 1916), published as War Department Document No. 515 by the War College Division of the General Staff to provide a supplement on aviation to the already existing general statement of United States military policy. This 18-page document presents a plea for expansion but does not develop aerial doctrine appreciably beyond the position enunciated earlier.

A brief but vigorous statement of a most advanced concept of air doctrine appears in *General Principles Underlying the Use of the Air Service in the Zone of the Advance, AEF* (Printing Office, AEF, 3 Oct. 1917). Despite its origin, this document does not bear the authenticating signature of either the Commander-in-chief of the AEF or his adjutant general. A title-page note, "These principles will be held in mind by all personnel . . . ," signed by Col. William Mitchell, gives a clue as to both the origin and unofficial status of this exposition of the functional division between strategic and tactical aviation.

A useful compilation of information on production is to be found in Col. G. W. Mixter and Lt. H. H. Emmons, *United States Army Aircraft Production Facts* (Washington, GPO, 1919). The two officers who compiled this report at the request of Assistant Secretary of War Crowell were both involved in activities of the Bureau of Aircraft Production during the war. Insofar as the report deals with statistics, the facts presented compare favorably with source materials in the papers of the BAP in the National Archives. But along with the facts the authors have injected much that is opinion which, coupled with the presence of many defensive omissions, requires one to utilize this document with caution. This compilation raises an interesting historical problem: just how "official" is such a publication? Although it was prepared "at the request of the Assistant Secretary of War" and issued by the Government Printing Office, the title page gives no indication of organizational responsibility. In the absence of any such indication, it seems imperative to assess the compilation as the product of the authors as individuals rather than as officials of an organization. In this particular instance alternate sources indicate the general correctness of the statistics presented. Nonetheless, the problem is worth mention because it is not unique. When dealing with government publications, it is well to remember that the imprint of the Government Printing Office is no guarantee of official sanction or approval or, for that matter, of accuracy.

For detailed information regarding the operations and activities of the Air Service, AEF, there is no publication comparable to the *Final Report of the Chief of Air Service, AEF*, originally submitted in March 1919 and subsequently reprinted in the *Air Service Information Circular*, Vol. 2, No. 180 (15 Feb. 1920). Although valuable as a compendium of factual information regarding aerial activities in the AEF, this report is surprising in its failure to derive significant lessons regarding doctrine in a coherent and codified form. The whole report is badly organized. Of a similar character is the *Final Report of Gen. John J. Pershing, Commander-in-chief, American Expeditionary Forces* (Washington, GPO, 1920), which is noteworthy, insofar as aviation is concerned, for its failure to indicate anything of significance about the air weapon.

A most useful compendium of information regarding materiel for the air arm in World War I is contained in Assistant Secretary of War Benedict Crowell, *America's Munitions; 1917–1918* (Washington, GPO, 1919). As in the case of the publication by Mixter and Emmons, the official character of the volume is open to question. The opinions expressed by the author (the volume was actually compiled by others) are not to be accepted as official declarations of policy even if the statistics on production are taken at face value. B. Crowell and R. F. Wilson, *The Armies of Industry* (New Haven, Yale University Press, 1921), is little more than a rewrite of the volume of 1919.

After the war the Historical Section of the Army War College prepared a study, *The Signal Corps and Air Service* (Washington, GPO, 1922), which appeared over the signature of the Chief of Staff as an approved monograph (No. 16 in the series at the Army War College) for use by officers of the army at large. The publication is of interest only as one to avoid. It is uncritical, badly organized, poorly written, difficult to use. As a monograph this work might properly be classified under secondary publications, but since its use is not recommended, it may not be out of place to mention it here with the source materials of the War Department.

For an indication of air doctrine taught after World War I as distinguished from the doctrine declared as official, Capt. E. L. Naiden's *Air Service* (Fort Leavenworth, Kans., The General Service Schools Press, 1920), a textbook of the Command and General Staff School, is probably the most reliable single index available. Subsequent texts of staff schools show little or no deviation in doctrine for a number of years. Both *Corps and Army Air Service* (Fort Leavenworth, Kans., The General Service Schools Press, 1922) and *Tactics and Technique of the Separate Branches* (Fort Leavenworth, Kans., The General Service Schools Press, 1924) repeat the tenets of the earlier text. The most noteworthy fact about postwar military publications on doctrine is the almost complete lack of interest in the subject on the part of the Air Service. Not until the thirties did texts begin to appear bearing the imprint of the air arm.

## Congressional Publications

Congressional publications provided a wealth of source material for the preparation of this study. Some of the more important are discussed in chronological order below. One might well question the propriety of using transcripts of Congressional hearings indiscriminately as a source of facts. That witnesses before Congressional committees frequently testify under oath does not, of course, alter the circumstance that testimony given months or even years after the events discussed is dubious, at best, regarding detail. Similarly, the Congressional imprint should not blind one to the character of the motives of witnesses. But despite special pleading and lapse of memory, Congressional hearings, especially those in connection with the War Department, have a high value as source materials. Officers frequently appear with true copies of reports and correspondence from official files which are reprinted in full. Judiciously used, these hearings can provide much in the way of sources. Congressional reports containing the findings or conclusions of such hearings are, of course, quite another matter.

One of the earliest Congressional publications on aviation, *Aeronautics in the Army*, Hearings before the House Military Affairs Committee on HR 5304, 63 Cong. 1 Sess., 12 Aug. 1913, demonstrates one of the points mentioned above. While the testimony of various witnesses reflects opinions, guesses, and bad memory, the published hearings include statistical data on the existing organization of the Signal Corps as submitted in a report of the Chief Signal Officer.

*Aircraft Production*, Hearings before a subcommittee of the Senate Military Affairs Committee, 65 Cong. 2 Sess., June to Aug. 1918, Vols. *1* and *2*, offers a wealth of material concerning the evolution of policies on the development of air materiel during the war. The final conclusions of these committee hearings, printed as *Aircraft Production in the United States*, Senate Report No. 555, 65 Cong. 2 Sess., 22 Aug. 1918, are of less value than the hearings themselves. The report not only is poorly organized but also fails to isolate and assess the critical difficulties of aircraft production during wartime. The Senate report is markedly inferior to that prepared by Hughes (see below), although it covers essentially the same area of interest.

The so-called Hughes Report was prepared at the President's request by the Department of Justice. The original report took the form of a letter from Charles Evans Hughes to the Attorney

General, 25 October 1918. The text of this letter was widely reprinted at the time. Although the text appearing in the periodical,
*Automotive Industries, the Automobile, 39*, No. 180 (31 Oct.
1918), 745 ff., has been used for convenience in preparing this
study, the report may also be found in the *Congressional Record,
57*, No. 26 (3 Jan. 1919), 1032–62. Even if one disregards the
conclusions of the report, the facts recited in this document were
of particular value in supplementing extant archival source materials. Many of the records quoted are no longer available. No
single document of the period gives such a comprehensive picture
of the difficulties of production during World War I.

The hearings of the Graham Committee of World War I, which
may be likened to the Truman Committee of World War II, contain two volumes devoted exclusively to aviation. This Congressional publication is entitled *War Expenditures*, Hearings, Select
Committee on Expenditures in the War Department, House of
Representatives, 66 Cong. 1 Sess., Subcommittee No. 1, Aviation,
Serial 2, Pts. 1–10, Vol. *1*, and Pts. 20–44, Vol. *2*, 1919. There is
a separate index volume. Although large portions of these hearings are concerned with fiscal irregularities, the printed testimony
of numerous officials contributes substantially to an understanding
of the process of evolving doctrine during the war. Valuable as
the hearings are, there is some justice to the minority report regarding the committee's conclusions published 2 March 1921.
These findings, the minority declared, were "biased, erroneous and
totally misleading" as well as "entirely useless for historical purposes." This judgment, of course, applies primarily to the committee's attempts at party recriminations and does not seriously
affect the value of much of the oral testimony and documentary
inclusions presented by officers of the air arm.

Of a somewhat less dubious character as sources are the Congressional papers published for the express purpose of presenting an
official record of status or achievement. In this category, a most
useful compilation is *Report on Aircraft Surveys*, House Document No. 621, 66 Cong. 2 Sess., 19 Jan. 1920. This publication
includes tabulations of all contracts for aircraft and engines between 6 April 1917 and 1 November 1919, showing numbers
ordered, numbers delivered, and money expended. No publication
of the War Department presents such a complete résumé.

After World War I there were numerous hearings and investigations, Congressional and otherwise, regarding the air arm, but
despite their number and frequency few add anything of signifi-

cance to the field of this particular study. *Pioneer Aviators*, Hearings before the House Committee on Military Affairs on HR 11273, 70 Cong. 1 Sess., 3 April 1928, presents a useful bibliography (see section on bibliography, above) and some charts and tables of information not found elsewhere in one codified form. The brief historical résumé of the air arm appearing in this publication must, however, be used with caution, if at all. This historical sketch is filled with typographical errors as well as downright misrepresentations of fact and is worth reading only as a brief orientation in the general pattern of military aviation in the United States.

## Miscellaneous Official Publications

Outside of documents from the War Department and from Congress, there are few official publications bearing directly upon the focus of interest in this study. The annual reports of the National Advisory Committee for Aeronautics (Washington, GPO, 1916 to date) are a notable exception. The reports for the years 1915–20 are of particular interest.

## Foreign Official Publications

No résumé of official publications would be complete without reference to the outstanding British publication on military aviation in World War I. Because of his untimely death, Sir Walter A. Raleigh (*The War in the Air* [Oxford, Clarendon Press, 1922]) wrote only the first volume of a projected series. However, H. A. Jones (*The War in the Air* [5 vols. *2–6* and an appendix volume; Oxford, Clarendon Press, 1928–37]) completed the series as planned by Raleigh. Although not actually a part of the British official history of the war, the series was written at the direction of the Committee of Imperial Defense and based on official documents. This study presents a history of the RAF and its predecessor agencies during World War I and was written from records of the Air Ministry collected by the Air Historical Section. The volumes represent a monumental compilation of information on the air war. Emphasis is on operations. Both administration and development of materiel are neglected by Raleigh who tends to dwell on personalities and to underplay if not to whitewash some of the organizational difficulties encountered in forming the Air Ministry. Jones makes some amends for these deficiencies in the subsequent volumes. Nevertheless, the whole question of production, research,

changes in design, and the problem of maintaining superiority of
performance is compressed into one or two chapters, very few pages
more than are devoted to the trivial aerial operations in South
Africa. Despite this obvious lack of proportion, these volumes are
an invaluable source in any study of the air weapon. Even if one
questions the validity of the authors' conclusions, the series is still
extremely valuable for the numerous appendices in each volume
and the separate appendix volume reprinting critical documents
on policy, tabulations on production, and the like.

Perhaps the most valuable contribution of the Raleigh-Jones
volumes lies in the documents reprinted at length to show the evolution of British concepts of doctrine on air power. The importance
of these documents is enhanced by the want of counterparts in the
United States where the problem of air doctrine was never given
the same degree of speculative interest it received in Great Britain.
The absence of any study in the United States comparable to the
Raleigh-Jones volumes testifies to this lack of interest. French
studies on the air arm tend to concentrate on operations rather
than upon the administration and development of materiel without exploring the obvious relationship of the two. An interesting
problem left unsolved is the transfer of leadership in the concept
of strategic bombardment from the French in 1917 to the British
in 1918. Official German studies on the air weapon were necessarily
impaired by postwar restrictions on the air arm. Such few publications as are available appear to emphasize operations rather
than administration. One of the more surprising revelations turned
up while preparing this study was the almost complete failure of
military officials in the United States to make use of foreign source
materials in writing on aerial warfare. Studies on aviation undertaken by students at the War College during the twenties appeared
to ignore foreign publications and reports from attachés, relying
instead upon dubious sources, in at least one instance the *Literary
Digest*.

## Secondary Publications

In the general category of secondary literature, as in the matter
of bibliographies mentioned above in Section I, two rather different types of works were of use—those dealing with the problem
of weapons in general and those dealing specifically with the aerial
weapon. Some of the more provocative studies falling within the
first of these two groups are discussed here:

Brig. Gen. C. D. Baker-Carr, *From Chauffeur to Brigadier* (London, E. Benn, 1930), is the memoir of an officer who played a leading role in popularizing the machine gun in the British army. Although the volume is primarily concerned with the machine gun, Baker-Carr's remarks on the process of "selling" the new weapon to higher authority have a general application. Major General E. D. Swinton's *Eyewitness* (London, Hodder and Stoughton, 1932) treats the same problem in the case of the tank. Admiral Sir Percy Scott, *Fifty Years in the Royal Navy* (New York, Geo. H. Doran, 1919), has many vitriolic pages dealing with this problem from the point of view of the navy. The thesis of Scott's work, the need for method in the development of weapons, is clearly stated in a brief introduction. This volume was suggested to the author by James Phinney Baxter III, president of Williams College, who has built a reputation on *The Introduction of the Ironclad Warship* (Cambridge, Harvard University Press, 1933). This book represents a staggering amount of research in French and British as well as native source materials but is strangely silent as to the administrative mechanisms behind the process of development. President Baxter's more recent book, *Scientists against Time* (Boston, Little, Brown, 1946), dealing with research on weapons in World War II, is far more appreciative of the all-important problem of administration. A whole chapter is devoted to the question of the relationship of strategy, or doctrine, and weapons. An American equivalent to Sir Percy Scott's book may be found in E. E. Morison, *Admiral Sims and the Modern American Navy* (Boston, Houghton Mifflin, 1942).

One of the most prolific students of the problem of weapons is the British officer, Maj. Gen. J. F. C. Fuller, whose writings are almost never absent from the pages of contemporary military journals. His *Armament and History* (New York, Scribner's, 1945), a study of the influence of armament on history, is but the most recent of a long series of studies appearing with almost perennial frequency since World War I. Although Fuller suffers from rushing into print and, as a consequence, frequently changes his opinions, his works are nevertheless rewarding to the student of armament. Even such lesser-known titles as *The Reformation of War* (New York, E. P. Dutton, 1923) and *The Foundations of the Science of War* (London, Hutchinson, 1926) contribute provoking analyses of the role of weapons in warfare and the influence of technology on doctrine.

B. H. Liddell Hart, another British author known for his popular writings on the military art in general, has contributed to the few available studies on weapons and doctrine. Liddell Hart, *The British Way in Warfare* (London, Faber and Faber, 1932), despite its misleading title, contains a series of essays on the impact of technology upon military policy. Lewis Mumford, author of *Technics and Civilization* (New York, Harcourt, Brace, 1934), as a nonmilitary student of technology and society discerns the influence of weapons on warfare and industrial economy even more fully than most military writers. Mumford's familiarity with European studies on the technology of war makes his analysis of particular significance since most studies of weapons and doctrine have tended to linger within national borders. Tom Wintringham, *The Story of Weapons and Tactics from Troy to Stalingrad* (Boston, Houghton Mifflin, 1943), which presents a survey of the relationship of weapons and doctrine in a popular vein, is useful for orientation despite its cursory treatment of the subject. A rare and unusually thoughtful volume on the relationship of weapons, politics, and industry is Sir J. Emerson Tennent's *The Story of the Guns* (London, Longmans, Green, 1864).

Without actually analyzing the cause of the difficulty, F. A. Shannon (*The Organization and Administration of the Union Army; 1861–1865* [Cleveland, Arthur H. Clark, 1928]) manages to give a vivid picture of the results obtained from inadequate organization for developing weapons. Volume *1* contains a chapter, "The Problem of Munitions," which is more valuable for the questions it raises than for the explanations it offers. For all its obvious deficiencies, this volume represents a useful case history illustrating the fundamentals of the relation between weapons and doctrine.

From the paucity of titles listed and the limited character of the contents discussed, it must be readily evident that the literature on the kinship of weapons and doctrine is extremely scanty. While this may not be particularly unexpected, the poverty of commentary on the aerial weapon in World War I is really surprising.

Several studies were particularly useful in exploring foreign experience with the aerial weapon. G. P. Neumann (one time major in the German air force), *The German Air Force in the Great War*, tr. J. E. Gurdon from 1920 Berlin ed. (London, Hodder and Stoughton, 1921), contains some interesting observations on doctrine. Somewhat narrower in scope is A. P. Voisin (onetime general in the French air force), *La Doctrine de l'aviation française de combat au cours de la guerre 1915–1918* (Paris, Berger-Levrault,

1932). This volume is a splendid example of classical French exposition. It is an analytical study which attempts to derive principles and doctrine from the lessons of experience. Although he is primarily concerned with the function of army-cooperation, by logical extension on experience, the author emphasizes the importance of strategic bombardment, "the dreaded instrument of the future." F. W. Lanchester, *Aircraft in Warfare* (New York, D. Appleton, 1917), illustrates the extensive consideration given to the question of doctrine in Britain; a preface by Maj. Gen. Sir David Henderson is especially provocative. John R. Cuneo has completed but two volumes of an extended study, *Winged Mars*, dealing with the employment of aircraft in warfare. The two volumes already available, *The German Air Weapon; 1870–1914* and *The Air Weapon; 1914–1916* (Harrisburg, Military Service Publishing Co., 1942–47), bring the account down through 1916. These volumes deal with the military air arm in France, Britain, and Germany.

No bibliography of studies on air doctrine would be complete, of course, without at least passing reference to the work of the Italian theorist on air power, Gen. Giulio Douhet, *The Command of the Air* (New York, Coward-McCann, 1942). The influence of Douhet on concepts of doctrine held by officers in the Air Service during the twenties was probably more extensive than is generally recognized. This impact, however, came largely in the period directly following the area of interest of this study. Like most students of air power, Douhet neglects the problems of administration behind materiel and is even curiously indifferent to the importance of superiority of design.

Secondary literature in the United States on the administration of developing air weapons during World War I is confined to a mere handful of titles. The best account of air materiel in the war is contained in Arthur Sweetser, *The American Air Service* (New York, D. Appleton, 1919). Although not formally documented, this book was compiled by an officer of the Air Service with access to official records. Written too soon after the war to have much perspective, it nevertheless presents an extremely useful résumé of the problems encountered in development and production. The author is apologetic and defensive regarding some of the more controversial failures of the air arm. Sweetser's volume, though it is useful in the absence of a better one, can by no stretch of the imagination be called critical or analytical. T. M. Knappen, *Wings of War* (New York, G. P. Putnam's Sons, 1920), describes itself as an account of aircraft invention, engineering, development, and production during the war. As this book was not based on records other than those generally available to the public, it has little to offer save a number of errors of fact and many oversimplifications. Of the too few volumes available in this field, this is one to avoid.

The only analytical study of administration in the AEF is H. A. Toulmin Jr., *Air Service, American Expeditionary Force, 1918* (New York, D. Van Nostrand, 1927), which sets out to prove, implicitly, that the Air Service, AEF, was utterly disorganized until the Coordination Staff, of which the author was a leading member, put matters right. Although the author reaches no very impressive conclusions and despite the fact that documentation is limited to internal evidence, this study presents an unusual picture of the problems encountered in evolving a staff. The analysis is the more interesting in that it comes from outside the professional air arm.

A useful supplement to Toulmin's study is to be found in E. S. Gorrell, *The Measure of America's World War Aeronautical Effort* (Northfield, Vt., Norwich University, 1940), which was originally delivered as Cabot Lecture No. 6 at Norwich University and subsequently published in book form. The statistical tables and tabular presentations included are based on the materials of the typescript history of the Air Service (in the National Archives) which Gorrell edited 20 years earlier. Judged for its statistical data, this volume might well have been cited as a primary source. However, the conclusions and random comments of the text are unquestionably hindsight and revised opinions. Cabot Lecture No. 7 in the Norwich Uniersity series, C. G. Grey, *History of Combat Airplanes* (Northfield, Vt., Norwich University, 1941), offers a good brief study of the outstanding types of aircraft in the war. The same author's *A History of the Air Ministry* (London, G. Allen and Unwin, 1940) is more significant. This volume is a chatty, journalistic study, undocumented but based on a lifetime of personal experience and wide association in aeronautical circles. As editor of an influential British periodical on aviation, the author participated actively in the events he discusses. Personalities are emphasized unduly.

For ready comparisons between the air arms of the United States and the United Kingdom, H. A. St. G. Saunders, *Per Ardua; The Rise of British Air Power, 1911–1939* (London, Oxford University Press, 1945), is a good brief survey although little more than a not too capable condensation of the Raleigh-Jones volumes.

This book contains rather shocking errors regarding technical details and emphasizes tactical encounters and personal narratives at the expense of critical analyses of some of the broader questions of air power. J. M. Spaight, *The Beginnings of Organized Air Power* (London, Longmans, Green, 1927), offers the only available comparison of the administrative systems in the air arms of Germany, France, the United Kingdom, and the United States. It is based almost entirely on newspaper accounts and secondary works.

Surprising as it may be, there is actually no official published history of the Air Service, AEF, save the totally inadequate 128-page study by the Historical Section of the War College mentioned above in the section on source materials in the War Department. Almost equally surprising is the absence of any really significant memoirs or biographies of important officials directly concerned with the air arm. There is, to be sure, a large number of memoirs of prominent leaders, political and military, of the period, but these devote a relatively small number of pages to aviation. One of the few biographies dealing with a prominent figure in aviation is H. G. Pearson, *A Businessman in Uniform* (New York, Duffield, 1923), a laudatory study of R. C. Bolling. Frederick Palmer, *Newton D. Baker: America at War* (2 vols.; New York, Dodd, Mead, 1931), is perhaps as useful as any biography of the period, but even though the author had access to Baker's personal files as well as to the cables of the War Department and other excellent sources, he contributes little not already presented by Sweetser. David Lloyd George's *War Memoirs* (6 vols.; Boston, Little, Brown, 1933–37), make for spirited reading but confirm one's distrust of retrospective views by elderly officials. General J. J. Pershing, *My Experiences in the World War* (2 vols.; New York, Frederick A. Stokes, 1931), is typical of the memoir school of writing. Reminiscent in character, chronological in organization, and undocumented, Pershing's volumes are of far less value to the student of military problems than they might have been had they analyzed functionally and critically some of the larger problems of command during the war. General Peyton C. March, *The Nation at War* (New York, Doubleday, Doran, 1932), appeared as something of a rebuttal to Pershing's work. One chapter is devoted to the Air Service. Since the whole volume is primarily concerned with special pleading for March *vs.* Pershing, it contributes little save to demonstrate the unreliability of memory. Some of March's contentions appear to be in flat contradiction of the evidence available in the files of the War Department. His own post-war volume scarcely does credit to his wartime career.

Even more disappointing than March's or Pershing's volumes is the work of Maj. Gen. M. M. Patrick, *The United States in the Air* (Garden City, N.Y., Doubleday, Doran, 1928). From a war-time Chief of the Air Service, AEF, one might well expect some-thing more than 70 pages of recollections concerning the AEF, in which the author makes no effort to analyze the fundamental prob-lems of the air arm in operation. Patrick dwells at length on per-sonalities and attempts no critical evaluation of the air arm at war.

Biographies of the majority of wartime leaders such as Marshal F. Foch, *The Memoirs of Marshal Foch*, tr. Col. I. B. Mott (Garden City, N.Y., Doubleday, Doran, 1931), proved of value only insofar as they mirrored an almost complete absence of in-terest in aviation. Many important biographies of leaders such as Maj. Gen. Sir H. M. Trenchard are yet to be written. Among the memoirs of the lesser figures, perhaps none is more interesting than A. H. G. Fokker and B. Gould, *Flying Dutchman* (New York, Henry Holt, 1931), with its significant account of the de-velopment of the synchronizing gear.

## PERIODICAL LITERATURE

There is a remarkable disparity between the wealth of popular periodical literature on aviation in World War I and the dearth of such materials bearing on the problem of weapons. Articles on air weapons in the journals on aviation of the period under dis-cussion are primarily concerned with the tactical aspects of op-erations on the one hand and the technical aspects of aircraft on the other. Few if any treat the administrative problems of develop-ing weapons. For orientation in the field, however, especially for items of interest regarding individual officers and officials who played important roles in wartime aviation, the periodical litera-ture is worth consulting. The British journals, *Aeroplane*, *Flight*, *Flying*, and *Aeronautics*, as well as the French *L'Aeronautique* are useful in addition to such publications in the United States as *Aviation*, *Aerial Age Weekly*, *Air Service Journal*, *U. S. Air Service*, *Aircraft Journal*, and *Air Power*, to name only the more outstanding publications. The *Journal* of the Franklin Institute has carried a number of items of subsidiary interest to this study. Present-day periodicals, for the most part, have little to offer. But

E. S. Gorrell, "What, No Airplanes?" *Journal of Air Law and Commerce* (Jan. 1941), is an important exception. Since he wrote as a retired colonel with a perspective of 20-odd years, Gorrell's conclusions may be suspect, but the factual data which he presents appear to be taken directly from the typescript of the History of the Air Service, AEF, in the National Archives.

Any attempt to catalogue the periodical literature of value in orienting the reader in the area of interest surrounding this study should not overlook the contribution of the recurrent NACA publication, *Bibliography of Aeronautics,* which has an extensive subject-author index in the field of aviation (see especially the titles listed under "military aeronautics"). In general, periodical literature contributed little to this study.

By way of summary, it would appear to be a reasonable evaluation to state that this study has been based almost entirely upon official records. The major portion of these records was either archival materials or publications of the War Department. The almost complete lack of secondary literature on the subject is perhaps the best index of the degree to which the question of the relationships of technological advance, military doctrine, and the development of weapons has been neglected by military officials.

# Index

Acceptance trials, *see* Service tests

Acceptances of aircraft, 172

Accessory equipment, 128

Administrative organization, 8, 15; British precedents, 80; British research, 30; criticized by British, 97; for decision making, 39, 68, 79, 119, 126, 142; for decisions on design, 70, 130, 153; for doctrine, 19, 33, 133, 169, 176 (*see also* General Staff); for information, 21, 78, 80, 82, 84, 126, 142; for innovations, 16, 29 (*see also* Weapons, procedures for development of); for liaison, 88, 91–101, 154, 156 (*see also* Administrative organization for information); Ordnance Board, 20; for procurement, 34; for production, 105, 119, 124, 155; for research, 103, 105–16, 155; Signal Corps before 1914, 28; survey of, 65

Admiralty, British, *see* British Admiralty

AEF, organization, 48; program, 48; ignores strategic aviation, 48; Board on aircraft, 87; relations with BAP, 87; cancels SPAD production, 126; programs, 136; doctrine, 139; favors strategic aviation, 146; First Army, 158; Second Army, 158; final report, 160; *see also* Chief, Air Service, AEF

Aerial torpedo, 171 n. 42

Aerial weapon, *see* Air power

Aerodynamics, 66

Aeronautical design, 119; *see also* Design

Aeronautical engineers, 103

Aeroplane, *see* Aircraft, and individual types by name

Agreement, DMA-BAP, 78; *see also* Rothermere-Foulois

Air Board, British, *see* British Air Board

Aircraft, 12; tactical, 28; commercial, 28; experimental, 65; British, *see* British aircraft; French, *see* SPAD; Italian, *see* Caproni; long-range, *see* Night-bomber; *see also* Air power; Development; Doctrine; Engines; Industry; Manufacturers; Production program; Tactical objectives; Weapons, mission of

Aircraft Board, formed, 68, 93

Aircraft suppliers, *see* BAP

Aircraft users, *see* DMA

Aircraft Production Board, 67; coordinates Army-Navy designs, 68, 108 n. 14, 142–3

Air "force," *see also* Strategic aviation; Strategic bombing

Airframe, 121

Air-ground relations, *see* Doctrine, tactical

Air Ministry, British, *see* British Air Ministry

Airplane Engineering Department, 76, 77, 155 n. 13

Airplane Engineering Division, 155 n. 13

Airplane Experimental Department, 104

Air power, early appreciation, 29; doubted in U.S., 31; minimized by General Staff, 41; influence of doctrine, 50; British, 54; strategic, 146; status of, 36; U.S. and foreign compared, 129, 131, 136, 149, 152 n. 7, 157, 162–3; *see also* Doctrine

Roman army, 10
Root, Elihu, 25, 28
Rothermere-Foulois Agreement, 144 n. 28
Royal Air Force, *see* RAF
Royal Aircraft Factory, 26, 123
Royal Flying Corps, 54, 150; establishes Experimental Branch, 30; air power doctrine, 159; *see also* RAF
Royal Tank Corps, 17
Royalties, 51, 52, 60
Royce, Lt. Col. R., 163 n. 18
Rumpler aircraft, 30
Russia, 30
Russo-Japanese War, 16
Ryan, J. D., 69, 88, 130 n. 32

Saber, 21
St. Cyr, 106
Sample aircraft, 55
San Diego aviation center, 33
Scammell, J. M., 13 n. 31
Science, *see* Administrative organization for research; Research
Science and Research Division, 112–4; moves to BAP, 114; duties redefined, 114–5; status, 116–7; disappears, 154
Scientific liaison, *see* Technical liaison
Scientific method, 15; *see also* Administrative organization
Scott, Adm. Sir Percy, 16–7
SE-5, 126; *see also* SPAD
Second Army, AEF, 158
Secrecy, 36
Secretary of War, *see* Baker, N. D.; War, Secretary of
Security, *see* Secrecy
Selection, *see* Decisions
Selection of types, 70; importance of, 62; static, 63
Selfridge, Lt. G. E., 27–8
Senate Military Affairs Committee, 119; *see also* Congress
Service aircraft, *see* DH-4
Service school, *see* Army War College; Command and General Staff School
Service test, 7, 22, 28–9, 32–4, 79, 153; tank, 17; demonstrate utility, 31; *see also* Reports, operational
Shannon, F. A., 10

Sharp, Ambassador W. G., 44
Short Brothers, Ltd., 30
Short-range bomber, *see* DH-4; Observation aircraft
Signal Corps, Aviation Section, 30, 103, 124, 150
Signal Corps, composition of, 27; aviation advances, 28; early program, 28; repeats Ordnance experience, 31; relation to Air Service, AEF, 46; invites Allied aid, 51; efforts at liaison, 52; clings to initial program, 59; relations with Government agencies, 65; inexperience, 66; failures in planning, 67; and Aircraft Production Board, 67; forms Equipment Division, 68; reorganized, 68; loses aviation role, 69; reliance upon missions, 86; ignorant of AEF structure, 86; Science and Research Division, 112–3; sheds aviation, 114; loses Science and Research Division, 114; doctrine undeveloped, 133; favors observation aircraft, 134; *see also* Aviation Section; Chief Signal Officer; Equipment Division; Science and Research Division
Signal Corps, Chief of, *see* Chief Signal Officer
Sikorsky, Igor, 30
Smithsonian Institution, 107
Somme, Battle of, 17
Spaatz, Maj. Carl, 157
SPAD aircraft, 62, 86, 125; selected, 60; obsolete, 61; canceled, 126
SPAD mission, 86
Spanish-American War, 25
Spaulding, Col. P. L., 88, 90
Spaulding Mission, 88–90, 96; adverse effects, 90
Specifications, aircraft, 28, 34, 40, 79, 104; initial army, 27
Springfield rifle, 25
Squier, Maj. G. O., 27, 35, 45
Staff studies, 161
Standard aircraft, 33
Standard Aircraft Corporation, 143
Standardization, 19, 33, 40, 119–20, 122, 141, 151; stultifies, 121, 123
Stanton, E. M., 9
Statistics, use of, 18, 21–2, 93; *see also*